Essentials of Tissue Banking

George Galea
Editor

Essentials of Tissue Banking

Springer

Editor
Dr. George Galea, MD, FRCP, FRC Path
Scottish National Blood Transfusion Service
Ellens Glen Road
Edinburgh EH17 7QT
United Kingdom
george.galea@nhs.net

ISBN 978-94-017-8502-0 ISBN 978-90-481-9142-0 (eBook)
DOI 10.1007/978-90-481-9142-0
Springer Dordrecht Heidelberg London New York

Softcover re-print of the Hardcover 1st edition 2010

Printed on acid-free paper

Springer is part of Springer Science+Business Media (www.springer.com)

Preface

A paradigm shift is taking in raising standards on all aspects of tissue banking from procurement to processing, from storage to clinical usage. A number of monographs have appeared giving guidance on GMP and quality systems that should be adhered to. There are also good guidelines on the significance of the new regulations that have been enacted.

However there is a significant lack of information on the scientific basis and rationale on why tissue banking practices have developed the way they have. Many of them have developed without much validation and have been in use for many years, copied from one tissue bank to another. There are some good articles on some scientific aspects of tissue banking practices, but they are in relatively old and in somewhat obscure journals. This book is intended to fill this gap by getting a series of eminent experts to each write a chapter or two on up to date tissue banking practices.

Moreover in the modern world of tissue banking, safety and quality of tissues have taken an increasingly important role. In fact these activities form the basis of many licensure activities that relate to tissue establishments. An attempt has been made to cover these aspects as well.

Although the legislation covers all types of cells, including stem cells, they have been deliberately omitted. The target audiences for stem cells are different and the set up is also generally quite different unlike cord blood, which is very similar to tissues and therefore included.

In order to limit the size of the book to a manageable level the authors have been only asked to cover the basis of current practices, rather than future developments, such as embryonic cell developments, tissue engineering and gene therapy, which are more cellular therapies than tissue banking, per se. Courses conferring certificates and diplomas have just started to be provided by various professional bodies. This is important because tissue banking is becoming a distinct modality for scientists who want to work in the field. I believe such an audience would be very keen to use this book. Also any scientist and medical person working in the field should be interested in it. It will also be useful for many surgeons who use tissues in their daily clinical practice.

I am very grateful to all the authors, who without exception willingly gave so much time, effort and energy in writing their chapters. They are without exception, leaders in their field. They have all produced up to date and state of the art chapters. They have done it in the style they felt was appropriate and no attempt was made to modify this. The key objective was to collate all this information in one place. I have learnt a lot whilst editing this work and I hope that the readers will be engrossed in reading this book as much as I have.

Edinburgh, Scotland George Galea

Contents

Contributors

W. John Armitage CTS Bristol Eye Bank, University of Bristol, Bristol Eye Hospital, Bristol BS1 2LX, UK, w.j.armitage@bristol.ac.uk

John Barbara NHS Blood and Transplant, Colindale Avenue, London NW9 5BG, UK, Marina.mobed@nhsbt.nhs.uk

Scott A. Brubaker American Association of Tissue Banks, McLean, VA 22101, USA, brubakers@aatb.org

Ted Eastlund Department of Pathology, University of New Mexico School of Medicine, Albuquerque, NM 87131, USA, deastlund@salud.unm.edu

Eliane Gluckman Eurocord, Hematology Department, APHP University Paris VII, Paris, France, eliane.gluckman@sls.aphp.fr

Ellen Heck Transplant Medical Services, UT Southwestern Medical Center, Dallas, TX 75390-9074, USA, Ellen.Heck@UTSouthwestern.edu

Artur Kaminski Department of Transplantology and Central Tissue Bank, Medical University of Warsaw, Warsaw 02-004, Poland, akamin@ib.amwaw.edu.pl

John N. Kearney Head of Tissue Services, NHSBT Tissue Services, Liverpool L24 8RB, UK, John.Kearney@nbs.nhs.uk

Alan Kitchen National Transfusion Microbiology Reference Laboratory, NHS Blood and Transplant, Colindale Avenue, London NW9 5BG, UK, alan.kitchen@nhsbt.nhs.uk

Linda Lodge Scottish National Blood Transfusion Service, Edinburgh EH17 7QT, Scotland, Linda.Lodge@nhs.net

Aurora Navarro Banc de Sang i Teixits, pg Vall d'Hebron, 119-129 08035, Barcelona, Spain, anavarro@bstcat.net

Robert Parker Heart Valve Bank, Brompton Hospital, Sydney Street, London SW3 6NP, UK, R.Parker@rbh.nthames.nhs.uk

David Pegg Department of Biology (Area 14), University of York, York YO10 5YW, UK, dep1@york.ac.uk

Waclaw Stachowicz Institute of Nuclear Chemistry and Technology, Warsaw 03-195, Poland, w.stachowicz@ichtj.waw.pl

Izabela Uhrynowska-Tyszkiewicz Department of Transplantology and Central Tissue Bank, Medical University of Warsaw, Warsaw 02-004, Poland, ityszk@gmail.com

Ruth Warwick Consultant Specialist for Tissue Services, NHSBT Tissue Services, Collindale Avenue, London NW9 5BG, UK, ruth.warwick@nhsbt.nhs.uk

Martell K. Winters Nelson Laboratories, Salt Lake City, UT 84123, USA, mwinters@nelsonlabs.com

Part I
Management of

Chapter 1
Live Donors of Tissue

Ruth Warwick

The Process of Making a Living Donation

Donations by living donors include bone, cord blood, amniotic membrane, ophthalmic limbal stem cells and heart valves. The latter may include heart lung block recipients who then donate their hearts. Allograft haemopoietic stem cell donors of bone marrow and peripheral blood stem cells are also clearly living donors but are not dealt with in this chapter because their donations are not banked but used directly after donation. Living tissue and cell donors dealt with in this chapter, and the circumstances of their procurement, are summarised in Table 1.1.

The process of living donation of tissues shares many characteristics with blood donation, including a number of ethical issues relating to consent, ownership, a duty of care to the donors by the tissue facility as well as selection and testing of donors. However, there are also many differences and, whereas blood donors are self-selected, living tissue donors make their donations in the context of an operative procedure which they undergo for their own benefit.

For living bone donors the context is a hip joint replacement operation and the potential donor is actively approached, either by the surgeon's team or by an employee of the tissue facility, to see if the individual is interested in donating their femoral head removed as part of their therapeutic hip replacement operation.

For amnion donation the potential donor is an obstetric patient undergoing elective caesarean section for delivery of her infant. In the case of a cord blood donor the potential donor is the maternal-infant pair donating cord blood after the delivery of the infant. In this case, who the actual donor is, is less clear than in femoral head donation in that the mother may provide consent and her own medical and behavioural history but this is only as a surrogate for her infant who, some may argue, is the actual donor, a situation which is not yet tested in the courts.

R. Warwick (✉)
Consultant Specialist for Tissue Services, NHSBT Tissue Services, Collindale Avenue, London NW9 5BG, UK
e-mail: ruth.warwick@nhsbt.nhs.uk

G. Galea (ed.), *Essentials of Tissue Banking*, DOI 10.1007/978-90-481-9142-0_1,

Table 1.1 Tissues and cells given by living donors

Tissue	Donor source	Operation	Donor interview staff	Procurement staff	Use of tissues
Bone	Patients with osteo arthritis of the hip requiring hip joint replacement	Hip replacement with a prosthetic joint	• Tissue establishment staff • Hospital staff from the surgical team under a formal contract with the tissue establishment	Surgeons in operating theatre	Hip revision surgery using impaction surgery
Amniotic membrane	Obstetric patients	Elective caesarean section	• Tissue establishment staff • Hospital staff based on a formal agreement • Between the tissue facility and hospital staff	Midwifery or obstetric staff	• Ophthalmic surgery applications as biological dressings particularly where the ocular surface is ulcerated as in chemical burns • Biological dressing over limbal stem cell replacement • Skin replacement
Cord blood	Unrelated donation (public bank)	Ideally after delivery of the placenta without interference with the delivery of the infant or the care of the mother and infant pair	• The model for obtaining consent and medical history varies between establishments • Ideally there is separation between the staff obtaining consent and the staff caring for the maternal infant pair	• Cord blood bank staff • Contracted arrangements with obstetric and midwifery staff • Contracted arrangements with venesection service providers	Bone marrow and immunological replacement for conditions such as • Haematological malignancy • Immunological deficiency • Metabolic disorders • Storage disorders • Abnormalities in haemopoietic blood system including thalassaemia and sickle cell disease • Postulated source of non-haemopoietic stem cells for regenerative medicine • Immunomodulatory activity from mysenchymal stem cells

Table 1.1 (continued)

Tissue	Donor source	Operation	Donor interview staff	Procurement staff	Use of tissues
	Related donation for high risk families	Ideally after delivery of the placenta without interference with the delivery of the infant or the care of the mother and infant pair In some cases there may be collection earlier in delivery on the basis of individual circumstance based on a risk assessment	Hospital obstetric staff or midwifes working to cord blood bank procedures	Models vary but include supervision by cord blood staff of: • local midwifery or obstetric staff using telephone support and documentation and kit provision	Bone marrow and immunological replacement for: • Haematological malignancy • Immunological deficiency • Metabolic disorders • Storage disorders Abnormalities in haemopoietic blood system including thalassaemia and sickle cell disease where there is an individual in the family of the index pregnancy and the new infant may be a match for the sick child
	Autologous or family donation in low risk families (commercial banking)	Ideally after delivery of the placenta without interference with the delivery of the infant or the care of the mother and infant pair	Models vary between the commercial organisations	Models vary and may use: • Contracted arrangements with obstetric and midwifery staff • Contracted venesection services • Ad hoc arrangements with partners of the donating infant. The latter is less than ideal as training of partners	• Very rarely used for autologous haemopoietic donation for leukaemia as donation may contain cells with predisposition to disease to be treated. • In case of inherited defects the autologous cord blood will contain the same defect. • May be suitable for sibling treatment of diseases usually treated with bone marrow transplanted but outside of

Table 1.1 (continued)

Tissue	Donor source	Operation	Donor interview staff	Procurement staff	Use of tissues
				for cord blood collection may be operationally problematic	high risk families with existing patients with such diseases the chance of such disease is low. • Potential for future use to treat degenerative diseases (still in the research phase).
Wharton's Jelly	Obstetric patients	Wharton's jelly surrounding cord (not cord blood) may be used to source mesenchymal stem cells	No operational therapeutic banks at present	No operational therapeutic banks at present	Contains high numbers of mysenchymal stem cells which may be feasible to use as a source of cells for immunomodulation or tissue engineering [46]
Heart valve	Patients undergoing heart lung block transplantation for cystic fibrosis where the heart may be unaffected	Explanted heart may be used as a source of heart valves	• Tissue establishment staff • Under contract with surgical team according to a clearly defined specification	Heart, heart/lung procured at time of explantation prior to heart lung transplant	Replacement of abnormal valves which may have been acquired or congenital abnormalities
Limbal Stem cell donation	• May be autologous donation where one eye is unaffected by limbal stem cell loss • May be a syngeneic (identical twin donation)	• Autologous procedure • an elective altruistic related	• Pre operative assessment of donor by surgical team • Pre-operative assessment of related donor by surgical team • Organ coordinator	Procured by ophthalmic surgeon and transplanted directly	• To replace ophthalmic limbal stem cells in cases of auto-immune disease affecting the eyes. • Chemical and thermal burns • Ocular pemphigoid

Table 1.1 (continued)

Tissue	Donor source	Operation	Donor interview staff	Procurement staff	Use of tissues
	• Unlikely to be a living unrelated donor as the chance of an unrelated match are low and because the discomfort and risk of the donation make it unlikely or unethical to request such type of donation				

The result of donation of bone, amnion and cord blood in the situation where the donor is also a patient undergoing a medical intervention for their own benefit means that the donor is not a volunteer to quite the same extent as a donor of blood and they may be concerned that the care they receive as a patient may be compromised in some way if they decline to be a donor. Of course it is best practice to ensure that there is no coercion in obtaining consent to donation and this is dealt with in detail elsewhere in this book. However, it is relevant to the special aspects of tissue and all donation from living patients that the perception of the donor of themselves as a patient may influence the giving of consent and the medical history.

The interview with the living donor of tissues may be undertaken, not directly by an employee of the tissue establishment, but by staff from the surgical or obstetric department undertaking the procurement of the donation and such staff need to be trained by the tissue establishment in all the relevant aspects of medical, behavioural and travel history taking and the consent discussion. This applies to both the initial training of staff undertaking interviews and also applies to continuing training and updating as processes and regulation changes. Audit, as a means to examine whether standards are adhered to, may identify areas needing additional training. New developments in the field may also result in the need for staff updates.

Circumstances of Living Donation: How Are Tissues from Live Donors Obtained?

It is necessary to be able to identify the potential donor for each type of donation and the circumstances around each type of donation are different.

Femoral Head Donation

In the case of femoral head donations, potential donors may be identified in the outpatient department when they attend for surgical and medical assessment prior to their surgery. For those hospitals which run preoperative assessment clinics, these clinics may be an ideal time to approach potential donors when they are counselled about their forthcoming operation. They can be asked whether they are interested in donating the femoral head which otherwise would be discarded. At this time they will not be receiving pre-operative medication or receiving post operative analgesia and therefore they should be able to make an informed decision, unclouded by medication. They may be nervous that their decision may influence their surgical procedure or its timing and staff must be sure not to inadvertently coerce patients to donate.

There are a number of models for obtaining consent from femoral head donors. They may be interviewed face to face by tissue establishment staff or by trained hospital staff working according to specified procedures. Hospital staff may be contracted to the tissue facility and trained to use formal questionnaires, agreed in advance and regularly updated using a document control system within a quality management system. They may be interviewed by telephone, by arrangement and with their prior consent, using the same questionnaire where the interview is documented using digital recording and where the telephone discussion becomes part of their donor record. The advantages of the latter model are considerable: the interview can be at a convenient time chosen by the donor in advance, the recording can be audited and used as a training tool and not only is the consent recorded but also the prior information provided to the donor. However such a system does require resource and infrastructure to support it. All methodologies which are employed must ensure confidentiality and the timing of the interview must allow for donor competency issues, avoid the effects of drugs and anaesthetics and language comprehension issues.

Patients need to be informed that, if they wish to make a bone donation, it will involve them also giving permission to be tested for microbiology markers for infection which, in the UK, includes hepatitis B and C, HIV, HTLV1, HTLVII and syphilis and that where the results are pertinent to their health that they will be informed of those results and be advised how they should be managed. There may be additional discretionary microbiological tests such as for malaria or Chagas' disease if the travel history indicates the need for this. Patients also need to be told that, in those tissue banks where living tissue donors are recalled and retested for the mandatory microbiological markers at 6 months post-donation, they would need to provide another blood sample as well as one at the time of donation or very near to their operation. Potential donors need to know that a comprehensive medical and behavioural history is needed to assess the suitability of the individual as a donor. The history taking is relatively time consuming and individuals should ideally be informed that all the questions are asked of every potential donor. They also need to be told that donations unsuitable for therapeutic use may still be suitable for research and development. Therapeutic use is most commonly in revision of hip replacement operations. Research may be in the public sector where the outcome of research using donations of tissue may be published. Development work within the tissue establishment can lead to improvements in processing or other aspects of tissue bank procedures. Research may also be possible in the commercial domain, including for drug discovery work, although this would not benefit the donor financially. All these aspects may result in potential donors wishing to decline the invitation to donate. For those donors who agree to allow donation then the first blood sample is commonly taken at the time of anaesthesia prior to the hip replacement operation. The second sample may be taken by the community doctor (primary care practitioner) 6 months post donation, then posted back to the testing laboratory or tissue facility. The European Commission (EC) Directive [1] allows Nucleic Acid Testing (NAT) at donation as an alternative to recall and repeat testing of the donor. Patients who make donations need to be informed that the process of donor selection and testing mean that not all donations can be used therapeutically [2].

In the operation theatres a member of staff trained in the procurement of bone will have responsibility for holding and maintaining a stock of sterile packaged pots, supplied by the tissue facility, to be used to contain the donated femoral head after its removal by the surgeon. The surgeon will remove an analyte validated to represent the donation, commonly bone chips from the femoral head, and place these into broth for aerobic and anaerobic bacteriology culture according to the requirements of the individual bank. All the documentation, bone pots and associated bone chips for culture and blood samples for microbiology testing will be labelled, according to preset procedures, with donation barcodes or other means of identifying the donor and samples to allow tracking from the donor with the associated samples and documentation and subsequently to recipients. Ideally the labelling will be compatible with testing laboratory equipment so that manual transcription of information about the donor is avoided and computer traceability is ensured. It is a requirement of the EC Tissue and Cell Directives that a European Coding system is developed [3].

In the case of femoral head donation there is a need for extensive interaction between tissue facility and outpatient and theatre orthopaedic staff with the need for training and audit and updates to staff training.

The general principles described for femoral head banking also apply to other living donations of tissue. Special considerations for other types of living donations follow.

Amniotic Membrane Donation

In the case of amnion the information about potential donors may be obtained through the obstetric lists of elective caesarean section deliveries. Elective procedures carry the advantage over emergency caesarean section because there is time to have a pre-consent discussion with the donor before commencement of labour and for consent to be given or declined in advance. The discussion prior to consent is similar in content to that described for bone donors with regard to selection by medical and behavioural history and for the need of microbiological testing, although the information about use of the donation will clearly differ from that for bone donors. Amniotic membrane is used in the UK mostly in ophthalmic surgery as a biological dressing [4] but is sometimes used elsewhere as a skin substitute for burns. Again there is a need for close co-operation between midwifery, obstetric out-patient and in-patient delivery suite staff.

Limbal Stem Cell Donation

Limbal stem cells are needed for ophthalmic epithelial tissue repair. They are situated in the Palisades of Vogt, which is found in the limbal region around the cornea. Limbal stem cells are capable of great proliferative potential and maintain normal corneal function. Chemical and thermal burns, Stevens-Johnson syndrome, and ocular pemphigoid can cause limbal stem cell failure. This may result in the need for limbal stem cell transplantation. If one eye is affected then the contralateral eye can donate limbal stem cells. If both eyes are affected then a donation from a member of the recipient's family, or from a deceased donor, may be used. Matched tissue type cells are best, so autologous cells are ideal. However stem cell expansion and innovative procurement of cells may be the way forward. Immunosuppression may improve graft outcome. Limbal stem cells and amniotic membrane grafts can be used together. This type of grafting is innovative so the good practice of evaluating outcomes is clearly indicated.

Cord Blood Donation

Cord blood banking can be unrelated and altruistic, where the donation is provided for anyone who needs the equivalent of a bone marrow transplant and who matches the donor. Such types of cord blood transplant require international collaboration to find matches.

There are also directed cord blood banks for families at high risk of a family member needing a cord blood or bone marrow transplant for an existing child or other family member with a relevant disease treatable by the equivalent of bone marrow transplant. This appears to be a very effective way of banking cord blood [5] and the donor selection procedures used in this circumstance must maximise safe donation, whilst recognising the advantages of a potential close family match. Such family donations avoid the delay involved in waiting for a matched sibling donor to grow large enough to donate bone marrow. Without pre-implantation tissue type matching of donor and recipient there is a 1 in 4 chance of an HLA match for individuals with the same two parents. For patients with a genetic disease the donor may be affected by the same disorder as the intended recipient.

The third type of cord blood bank are banks run by the commercial sector for families wishing to bank cord blood for sole use by their own family members. These types of cord blood unit are rarely used and are banked on the presumption that cells within them will be able to be used for regenerative medicine purposes in the future, especially for the treatment of degenerative diseases. Some banks undertaking this type of banking run on a public/private partnership basis in order to provide for the common good whilst respecting the wish of some families to undertake privately banked cord blood storage.

Models for collection of unrelated cord blood vary, with some centres interviewing the mother using cord blood bank staff and others using midwifery staff. Some collect the cord blood whilst the placenta is *in-utero* and others after delivery of the placenta. Guidance from the Royal College of Obstetricians [6] recommends that cord blood collection should not interfere with the normal delivery of the child and care of the maternal infant pair.

All models require processes which include careful and thorough review of the donor's history; donor microbiological testing and aseptic retrieval of donated material. These reviews of the donor are complicated because assessment of the mother is only an indirect indicator of disease in the infant. Also, because the cord blood may be stored for decades before it is selected for a recipient, there is the chance that new tests may be pertinent at the time when the cord blood is chosen for transplantation and so it is important that there is storage of donors' analytes for future testing.

Donor interview, by a trained member of staff, must include, as with femoral head donors, medical, behavioural and travel history, but for cord blood donors an ethnic history is also useful. Donor selection will reduce the risk of transmission of either genetic or infectious disease. Neonatal haemoglobinopathy screening is undertaken by some cord blood banks, which target donors from susceptible ethnic minority groups in order to identify haemoglobin abnormalities not recognisable at birth, such as sickle-cell disease which could be transmitted to a recipient. Sickle cell trait has been transmitted [7] without clinical consequence, but transmission of sickle cell disease could result in the recipient developing severe disease. Cytogenetic reports, when available, may identify genetic abnormalities in donor infants and such services are sometimes provided by public authorities. Information through health professionals including the primary care practitioner of the mother, as well as mothers themselves, may be invited to facilitate reporting of medical complications relating to infectious, genetic or malignant disease identified after donation.

Thorough cleansing of the umbilical cord prior to aseptic collection and use of a dedicated clean collection room, plus closed system processing where possible, or the use of controlled environment and laminar flow hoods will limit aerobic and anaerobic bacterial contamination of the donated cord blood if closed system processing is not used.

Donor Selection: Evidence Base and General Considerations

Transmissions of disease by organs, tissues and cells is well documented [8] and includes examples of both infectious diseases and non-infectious disease transmission, the latter including malignancy, auto-immune or other disorders. Red cell alloimmunisation is not strictly a disease transmission but can be a complication of transplantation. Transmission of infections by tissue transplantation has been reviewed in the United States [9] on the base of a denominator of one and a half million tissue transplants each year. The review examined reports received by the Food and Drugs Administration from 2001 to 2004 and concluded that improvements in adverse reaction reporting will improve the ability of tissue establishments to undertake their activities and the safety of tissues for transplantation. The EC Directive [10] requires that adverse events and reactions are reported to the European Member States Competent Authorities and subsequent central collation of this has been further developed by a European project called EUSTITE (http://www.eustite.org/.) The probability of blood-borne virus in tissue donors in the US has been documented [11] and although prevalence rates were lower than in the general population, they were higher than among first time blood donors. Brant found that the prevalence of infection low among English tissue donors, but the risk was higher than that among new blood donors [12].

Safety measures include avoiding disease transmission to the tissue or cell recipient by preventing contamination of the donation at procurement, processing or storage, by selection of the donor, by testing the donor for blood-borne infections and processing the donation and, where possible, to de-contaminate or sterilise it. This chapter mainly addresses aspects of donor selection.

The donor interview provides the first cut exclusion of potentially inappropriate donors. There are differences between donations of viable cells and tissues, and those that are non-viable and their potential for disease transmission. There are also some recipient factors which can affect disease transmissibility. Immuno-competent and non-immuno-competent recipients differ in their susceptibility to donor derived disease. There are implications for other types of disease transmissions, such teratogenicity via drugs taken by the donor, relevant to female recipients of child bearing potential, and there is the need for matching of RhD red cell antigens when issuing femoral head donations containing blood or marrow. This is also relevant for cord blood recipients. Donors of cord blood may have a condition transferable by transplantation to the recipient of the cord blood. The implications of new donor selection criteria for tissue and cells already in inventory and selected according to obsolete donor selection and testing needs to be included in tissue and cell facility policies concerning review of stock inventory. Such policies can usefully include a risk assessment based approach on how to handle changes in selection criteria and the impact on donations already in inventory. This chapter uses examples of donor selection criteria for living donors to examine the evidence base for tissue or cell donor selection and highlights where the evidence behind such criteria could be enhanced. Of course there are many exclusion criteria but just a few of these are used as an example to illustrate the principles.

Exclusion of Living Donors of Bone

In the National Health Service Blood and Transplant (NHSBT), Tissue Services (TS) operates a living donor and deceased donor tissue banking programme. The living bone donor programme operates in collaboration with orthopaedic departments to collect surgically removed femoral heads. Organisations vary in their criteria for donor selection and an example of such criteria is that provided in the UK Blood Services Tissue Donor Selection Guidelines (www.transfusionguidelines.org.uk) and Guidelines for the UK Blood Transfusion Services [13]. The former gives exclusion criteria, such as autoimmune and infectious disorders, high-risk behaviour and malignant conditions. No age limit is applied for surgical donors of femoral heads but known osteoporosis, or other conditions providing high risk for osteoporosis are exclusions.

A survey [2] showed that the commonest exclusions from femoral donors are reported to be pre-existing bone or joint conditions and malignancy. The reported exclusions of donors of femoral heads for "malignancy" represent active malignant disease and non-metastatic disease or individuals in remission for some years. On the other hand some donors with no reported history of malignancy may have occult malignant disease, sometimes detectable in donated bone. Cases of low grade lymphoma were revealed in retrieved bone allografts [14] and evidence of malignant lymphoma and low grade chondrosarcoma were found in femoral heads otherwise considered suitable for donation [15]. Whilst transmission of malignancy is well documented for organs, the risk of transmission of malignant disease from tissues is remote, with only two reported cases where transmission occurred from a histiocytoma and a colonic carcinoma through direct tissue needle stick injury to the surgeon from the patient [16, 17]. In each of these cases the tissue was not bone and the tissue was not processed or frozen, unlike tissue used for bone banking and intended for transplantation.

Development of additional deferral criteria, aimed at reducing the risk of transmission of disease in general and, in the UK, of vCJD, will further reduce the donor base. In the UK femoral heads found suitable for clinical use are either issued as unprocessed fresh frozen donations containing blood and marrow, or they may undergo gamma irradiation. If new methods of processing of individual single FHs, including cell removal and sterilisation become available [18], it could be feasible to reconsider existing exclusions relating to donor malignancy and some auto-immune conditions. Frozen femoral heads have transmitted HTLV [19] showing that freezing does not remove the risk of transmission for cell dependent viruses like HTLV. However, freeze-thawing, hot hypotonic washing, ethanol and peroxide washing, re-freezing, freeze drying and 25 kGy gamma irradiation and ethylene oxide treatment are much less hospitable to cells and presumably therefore also other less kind environments to tissues. Peracetic acid also renders cells and viruses non-viable [20, 21]. Validation would be required to establish that such processing would remove the chance of survival of malignant cells or their causative agents from being transmissible to recipients of such processed bone.

The Commission Directive, 2006/23/EC requires that malignancy, with a few exceptions, be considered a contraindication to donation. However, it can be argued that donor acceptance and deferral criteria should be risk assessment based on evidence of transmissibility. Processing of tissues validated to remove cells may be a means to accepting donations from individuals with a history of malignancy or auto-immune conditions not affecting the specific tissue to be retrieved. Donor and recipient histo-incompatibility, plus recipient immune competence further reduce the chance of survival of malignant clones in the recipient and, together with the removal of oncogenic viruses, means that deferral for malignancy for bone donation could be reviewed on the basis of these combined factors. A firm evidence base is needed if standards for exclusion are to be effective in ensuring quality and safety of donated tissue, whilst preventing arbitrary exclusion of some tissue donors.

Other Examples of Selection of Living Tissue Donors to Reduce Risk of Disease Transmission

Blood and tissue donors are closely questioned about their behaviour to assess the risk of carriage of blood or tissue borne viruses. Questions include whether an individual has had exposure through medical intervention and lifestyles which are associated with a higher incidence of certain infections than other sectors of the community. Lifestyle risks include tattooing, sexual practices and non-prescription drug taking. There are many countries which require some type of registration of tattoo parlours or acupuncture facilities and personnel and that makes it easier to exclude potential donors who have had procedures in non-registered facilities where the chance of re-use of needles or tattoo ink cannot be excluded. Another method of dealing with such risk is to defer the donation for 12 months after the procedure to ensure that donors are not in the serological window period of infection. Alternatively deferral for 6 months may be employed if validated NAT is available at donation and hepatitis B anti core testing is performed to exclude donors infected with blood borne viruses in the serological window phase [13].

Obtaining behavioural history relating to tattoos and other procedures with similar risks is easier in the context of a first hand history from the donor, such as a femoral head donor, than in the context of a deceased donor where the history is provided by the next of kin or other family member. If a country does not have a licensing or registration system for tattoo facilities, investigation of facilities on an ad-hoc basis may be too time consuming to allow the tissue facility to be sure that the processes used were safe. When the European surveillance and vigilance schemes [22] for the reporting of community infections are routinely providing data then trends will be apparent and risks more easily identified so that the criteria for selecting donors will be more effectively evidence based. Such schemes will be able to identify risk areas and activities and differences between countries and within them.

Donor selection however must always be balanced carefully to ensure that selection and deferrals are proportionate or the availability of tissues for allografting will not be sufficient for patients' needs. It is also clearly a surgeon's responsibility, when using tissue for allografting, to ensure that that use of tissues is appropriate and that other approaches to treatment have been considered.

Males Who Have Had Sex with Males (MSM)

More traditional concerns relate to individuals in groups shown to have a high risk of carrying blood-borne viruses, such as men who have ever had sex with men (known as MSM). In some, but not all countries, MSM remain excluded for life from becoming donors of blood or tissues. It has been suggested that this criterion is too strict and that criteria for deferral should be consistent with rules for acceptance of other potential donors at increased risk for sexual transmission of blood-bourne diseases. The risks associated with MSM are much higher with regards to blood-borne viruses like HIV than the general population [23]. Risks also vary geographically and therefore donor acceptance criteria may also need to vary between countries. The medical and behavioural history is critical in assessing donor risk, whether it is for needle stick injuries, or for MSM. This is all in the context of sensitive microbiology assays using NAT for blood-borne viruses, an attitude to achieving zero risk and a limited supply of allograft tissues.

Selection of Cord Blood Donors

Microbiological Screening of the Mother and Cord

HIV and Hepatitis B and C

From 1988 to 1996 HIV infection rates rose six-fold in the heterosexual population [24] leading to concerns that a maternal donor may be in the HIV seronegative window period. In the first instance, the interview with the mother during consent may identify behaviour known to increase the risk of HIV infection. HBV has been transmitted by bone marrow transplantation [25]. Testing solely for HBsAg leaves a small risk of HBV infectious donations because of the possibility of a mother having antibody to HBV core (anti-HBc) as the only marker of potential HBV infectivity. However, the risk of such a mother being infectious to her child at birth is low and the cord blood is unlikely to be infectious. HBsAg is usually present in the patient's circulation within 1 month of exposure to infection. In some Asian and Afro-Caribbean populations, where HBV infection commonly acquired in infancy, often from an infectious HBsAg positive mother who is also HBeAg positive, chronic HBV infection is relatively common. Anti-HBc (together with anti-HBs) may be present in 50–80% of potential donors from endemic areas. If anti-HBc screening was used in the absence of anti-HBs testing, it would result in the unnecessary loss of donations which are not actually infectious [26].

Hepatitis C has been transmitted by blood and frozen bone, and it is likely that it could be transmitted by cord blood because there is between 6.2 and 10% transmission from mother to infant [27]. Additional testing, using high sensitivity techniques at the time of selection of a cord blood unit for transplant, will also reduce microbiological transmission risk if undertaken in addition to the mandatory microbiological markers performed at the time of donation.

Donor recruitment requires that information be provided to mothers during the antenatal period – leaflets, videos, posters and/or presentations at parent classes and in the appropriate languages for the potential donors, particularly if the obstetric unit where collection of cord blood is undertaken provides services to a multi-ethnic population. At the initial donor interview, the mother provides written consent, medical, behavioural, ethnic and travel history. Blood samples from the mother may be taken at the time of interview or at the time of cord blood collection. Collection either involves the placenta being passed to a member of staff, the umbilical cord cleaned with venesection of the umbilical cord using a needle attached to a closed collection bag or in some centres the cord blood may be collected whilst the placenta is still *in-utero*. At the time of cord blood collection, samples from the cord blood may be used for a variety of analyses; cell phenotyping, tissue typing and bacteriology screening. Analytes need to be archived for future testing. Samples from the mother and cord blood plasma can facilitate mandatory microbiology screening and for CMV, whilst DNA and plasma can be archived for future testing at the time of issue. Cord blood banks have varying procedures for banking collected units, depending on the volume or cell count.

Some agents may cause not only the pathognomic disease by which they are primarily known, but also other symptom complexes where the association is less well recognised. Iritis and arthralgia may be caused by HTLV but this association with HTLV may not be widely appreciated [28–31]. Sarcoidosis may be associated with human herpes virus 8 (HHV-8) infection [32].

Regamey described the transmission of HHV-8 infection from renal transplant donors to recipients [33]. However, [34] in the context of renal transplantation, transplant physicians should seriously consider HHV-8 infection in the donor population prior to transplantation. If HHV-8 in the moderately immuno-suppressed renal recipient population poses a significant problem, it is likely that in the more profoundly immuno-suppressed bone marrow or cord blood stem cell recipient, the problem may be greater. There are no screening programmes for HHV-8, and the specificity of tests currently available is uncertain. False positive tests would result in losses from the donor pool and health concerns for candidate donors. The association of HHV-8, with subsequent development of Kaposi's sarcoma, in profoundly immuno-suppressed individuals warrants consideration [33].

Not all diseases are subject to vertical transmission during pregnancy or delivery but vertical HIV transmission is well documented [35] and behavioural risk in the mother may be significant in assessing the safety of her infant's cord blood for transplantation. It is less clear whether maternal malignant disease is a risk to the cord blood recipient. There is a balance between ensuring safety, whilst ensuring sufficiency of donation. Donor derived malignant disease transmission transplantation has been reviewed recently [36].

Other Types of Potential Disease Transmission by Cord Blood

Trafficking of cells between the mother and the fetus occurs [37] and foetal cells in the maternal circulation may be long lived. Male foetal cells persist in maternal blood for decades after delivery [38]. Fetal DNA and cells have been identified in skin lesions from women with systemic sclerosis [39]. This shows the potential capacity of fetal cells to engraft against the maternal HLA barrier [40]. Maternal lymphocytes may survive in the infant as up to 4% of the total for up to 5 years following maternal in utero transfusions [41]. Some fetal cells are destroyed in the maternal circulation [42] but there has been a case of vertical transmission of lymphoma from mother to child, where both died of the lymphoma [43].

If there is a strong family history of malignancy of early onset, there is the potential for transfer of the predisposition to the cord blood recipient. This makes the family medical history taking very pertinent in making the decisions whether to transplant from a potentially affected cord blood donor in the context of availability of alternative donors. This is a decision for the transplant physician in collaboration with the cord blood bank medical director. Such information is also pertinent in the evaluation of second malignancies in the recipient transplant populations where second malignancies are not uncommon [36]. Other genetic risk markers could become the subject of future screening programmes. DNA archives are commonly maintained by cord blood banks. For example, acute lymphoblastic leukaemia (ALL) has been shown to be associated with an abnormal fusion protein that may be present before birth and precede the development of disease by several years. There has been a case of acute lymphoblastic leukaemia following cord blood transplantation, where the donor was known to be healthy at 7 years of age [44]. There have been 4 cases of donor cell derived acute myeloid leukaemia following unrelated cord blood transplantation where the donors were well 6–12 months after birth [45]. These cases illustrate unique ethical issues in cord blood banks.

Genetic Disease

Sickle cell trait [7] is one of a number of genetic disorders which can potentially be transmitted by cord blood transplantation. The potential exists for haemoglobinopathies, thalassaemia, and immune deficiencies, Fanconi's anaemia and the metabolic diseases for transplant transmission.

Travel History

Ethnic minorities may come from areas where viral infections are endemic and relevant to transplantation. Risk from semi-immune donors, born in a malarial endemic area, following childhood residence in such an area, especially with return to those areas in the previous 3 years may be relevant.

Bacterial Risk

Maternal puerperal pyrexia or prolonged rupture of the membranes may point to a risk of bacterial infection, as might infants with possible congenital bacterial infection associated with amnionitis. Frankly bacteraemic cord blood donations would be detected by positive cord blood cultures. Bacterial risks may be associated with the mother, the child, the environment or the staff who undertake collection, processing, cryopreservation or storage of the cord blood. To minimise bacteriological contamination from maternal infection, the mothers of cord blood donors should ideally be selected on the basis of having undergone a normal pregnancy, whether delivery is vaginal or by caesarean section.

References

1. Commission Directive 2006/17/EC of 8 February 2006 implementing Directive 2004/23/EC of the European Parliament and of the Council as regards certain technical requirements for the donation, procurement and testing of human tissues and cells. Official Journal of the European Union L 38/40 09/02/2006.

2. Pink F, Warwick RM, Purkis J, Pearson J (2006) Donor exclusion in the national blood service tissue services living bone donor programme. Cell Tissue Bank 7(1):11–21. DOI: 10.1007/s10561-005-2362-2
3. Commission Directive 2006/86/EC of 24 October 2006 implementing Directive 2004/23/EC of the European Parliament and of the Council as regards traceability requirements, notification of serious adverse reactions and events, and certain technical requirements for the coding, processing, preservation, storage and distribution of human tissues and cells. Official Journal of the European Union L 294/32 25/10/2006.
4. Saw VP, Minassian D, Dart JK, Ramsay A, Henderson H, Poniatowski S, Warwick RM, Cabral S, the Amniotic Membrane Tissue User Group (AMTUG) (2007) Amniotic membrane transplantation for ocular disease: a review of the first 233 cases from the UK User Group. Br J Ophthalmol 91:1042–1047
5. Smythe J, Armitage S, McDonald D, Pamphilon D, Gutteridge M, Brown J, Green A, Brown C, Warwick RM, Lankester A, Fehily D, Contreras M, Navarrete C, Watt SM (2007) Directed sibling cord blood banking for transplantation: the 10-Year experience in the national blood service in England. Stem Cells 25:2087–2093
6. Royal College of Obstetricians and Gynaecologists Scientific Advisory Committee, Opinion Paper 2, June 2006
7. Ruiz-Arguelles GJ, Reyes-Nunez V, Garces-Eisele J, Warwick RM, McKenna L, Ruiz-Reyes G, Granados J, Mercado-Diaz MA (2005) Acquired haemoglobin S trait in an adult patient with secondary acute myelogenous leukemia allografted with matched unrelated umbilical Cord blood cells. Haema 8(3):492–496
8. Eastlund T, Strong DM (2004) Infectious disease transmission through tissue transplantation. In: Phillips GO, Kearney JN, Strong DM, Von Verson R, Nather A (eds) Advances in tissue banking, vol 7. Scientific Publishing, Singapore.
9. Wang S, Zinderman C, Wise R, Braun M (2007) Infections and human tissue transplants: Review of FDA MedWatch reports 2001-2004. Cell Tissue Bank 8(3):211–219. Feb 2007. (E-pub. DOI 10.1007/S10561-007-9034-3)
10. Directive 2004/23/EC of the European Parliament and of the Council of 31 March 2004 on setting standards for quality and safety in the donation, procurement, processing, preservation, storage and distribution of human tissues and cells. Official Journal of the European Union L 102/48 07/04/2004.
11. Zou S., Dodd RY, Stramer SL, Strong DM for the Tissue Safety Study Group (2004) Probability of Viremia with HBV, HCV, HIV, and HTLV among tissue donors in the United States. N Engl J Med 351(8):751–759
12. Brant LJ, Davison KL (2008) Infections detected in English surgical bone and deceased donors (2001–2006) and estimated risk of undetected hepatitis B and hepatitis C virus. Vox Sang 95:272–279
13. Guidelines for the Blood Transfusion Services in the United Kingdom (2005) 7th edn. TSO (The Stationary Office), London. PO Box 29, Norwick, NR31 1GN www.tsoshop.co.uk
14. Sugihara S, Van Ginkel AD, Jiya TU, Van Royen BJ, Van Diest PJ, Wuisman PIJM (1999) Histopathology of retrieved allografts of the femoral head. J Bone Joint Surg 8(B):336–341
15. Palmer SH, Gibbons CLMH, Athansou NA (1999) The pathology of bone allograft. J Bone Joint Surg 81(B):333–335
16. Gartner HV, Seidl C, Luckenbach C, Schumm, G, Seifried E, Ritter H, Bültmann B (1996) Brief report: genetic analysis of a sarcoma accidentally transplanted from a patient to a surgeon. N Engl J Med 335:1494–1497.
17. Gugel EA, Sanders ME (1986) Needle-stick transmission of human colonic adenocarcinoma. N Engl J Med 315:1487
18. Lomas R, Drummond O, Kearney J.N (2000) Processing of whole femoral head allografts: a method for improving clinical efficacy and safety. CATB 1:193–200
19. Sanzen L, Carlsson A (1997) Transmission of human T-cell lymphotrophic virus type 1 by a deep-frozen bone allograft. Acta Orthop Scand 68(1):72–74

20. Pruss A, Kao M, von Verson R, Pauli G (1999) Virus safety of avital bone tissue transplants: evaluation of sterilization steps of spongiosa cuboids using a peracetic acid methanol mixture. Biologicals 27:195–201
21. Pruss A, Baumann B, Seibold M, Kao M, Tinteinot K, von Verson R, Radtke H, Dorner T, Pauli G, Gobel UB (2001) Validation of the sterilization procedure of allogeneic avital bone transplants using peracetic ethanol. Biologicals 29:59–66
22. Fehily D, Delvecchio C, DiCiacco P, Femaro C (2007) The EUSTITE Project: working towards harmonised implementation of European regulation of tissue and cells. Organs Tissues Cells 10(1):31–36.
23. Soldan K, Sinka K (2003) Evaluation of the de-selection of men who have had sex with men from blood donation in England. Vox Sang 84:265–273
24. Nicoll A, McGarrigle C, Brady AR, Ades AE, Tookey P, Duong T, Mortimer J, Cliffe S, Goldberg D, Tappin D, Hudson C, Peckham C (1998) Epidemiology and detection of HIV-1 among pregnant women in the United Kingdom: results from national surveillance 1988-96. BMJ 316(7127):253–258
25. Tedder RS, Zuckerman MA, Goldstone AH, Hawkins AE, Fielding A, Briggs EM, Irwin D, Blair S, Gorman AM, Patterson KG, Linch DC, Heptonstall J, Brink NS (1995) Hepatitis B transmission from contaminated cryopreservation tank. Lancet 346:137–140
26. Ohto H, Terazawa S, Sasaki N, Hino K, Ishiwata C, Kako M, Ujiie N, Endo C, Matsui A, Okamoto H, Mishiro S, for The Vertical Transmission of Hepatitis C Virus Collaborative Study Group (1994) Transmission of hepatitis virus from mothers to infants. N Engl J Med 330:744–750
27. Dore GJ, Kaldor JM, McCaughan W (1997) Systemic review of role of polymerase chain reaction in defining infectiousness among people infected with hepatitis C virus. BMJ 315:333–337
28. Sullivan MT, Williams AE, Fang C, Grandinetti T, Poieszbj, Ehrlich GD (1991) The American Red Cross HTLV-I/II collaborative study. Transmission of human T-lymphotropic virus types I and II by blood transfusion. A retrospective study of recipients of blood components (1993 through 1988). Arch Intern Med 151:2043–2048
29. Okochi K, Sato H, Hinuma Y (1984) A retrospective study on transmission of adult T cell leukaemia virus by blood transfusion: seroconversion in recipients. Vox Sang 46:245–253
30. Nightingale S, Orton D, Ratcliffe D, Skidmore S, Tosswill J, Desselberger U (1993) Antenatal survey of seroprevalence of HTLV-I infections in the West Midlands, England. Epidemiol Infect 110:379–387
31. Tosswill JHC, Ades AE, Peckham C, Mortimer PP, Weber JN (1990) Infection with human T cell leukaemia/lymphoma virus type I in patients attending an antenatal clinic in London. BMJ 301:95–96
32. Luca di A, Piattelli A, Artese L, Favia G, Patel S, Saunders N, Porter SR, Scully CM, Ngui SL, Teo CG (1997) Human herpesvirus 8 variants in sarcoid tissues. Lancet 350:1655–1661
33. Regamey N, Tamm M, Wernli M, Witschi A, Thiel G, Cathomas G, Erb P (1998) Transmission of herpesvirus 8 infection from renal transplant donors to recipients. N Engl J Med 339: 1358–1363
34. Ho M (1998) Human herpes virus 8-Let the transplant physician beware. N Engl J Med 339:1391–1392
35. Sperling RS, Shapiro DE, Coombs RW, Todd JA, Herman SA, McSherry GD, O'Sullivan MJ, Van Dyke RB, Jiminez E, Rouzioux C, Flynn PM, Sullivan JL (1996) Maternal viral load, Zidovidine treatment and the risk of transmission of HIV-1 from mother to infant. N Engl J Med 335:1621–1629
36. Ghandi MJ, Strong, DM (2007) Donor derived malignancy following transplantation: a review. CATB 8(4):267–286, Epub 2007 Apr 12
37. Lo YM, Lo ES, Watson N, Noakes L, Sargent IL, Thilaganethan B, Wainscoat JS (1996) Two way cell traffic between mother and fetus: biologic and clinical implications. Blood 88: 4390–4395

38. Bianchi DW, Zickwolf GK, Weil GJ, Sylvester S, DeMaria MA (1996) Male progenitor cells persist in maternal blood for as long as 27 years postpartum. Proc Natl Acad Sci USA 93: 705–708
39. Artlett CM, Smith JB, Jimenez SA (1998) Identification of foetal DNA and Cells in skin lesions from women with systemic sclerosis. N Engl J Med 338:1186–1191
40. Nelson JL, Furst DE, Maloney, Gooley T, Evans PC, Smith A, Bean MA, Ober C, Bianchi DW (1998) Microchimerism and HLA compatible relationships of pregnancy in scleroderma. Lancet 351:559–562
41. Hutchinson DL, Turner JH, Schlesinger ER (1971) Persistence of donor cells in neonates after foetal and exchange transfusion. Am J Obstet Gynaecol 109:281–284
42. Bonney EA, Matzinger P (1997)_ The maternal immune system's interaction with circulating foetal cells. J Immunol 158:40–47
43. Catlin EA, Roberts JD, Rodrigo E, Preffer FI, Ferry JA, Kelliher AS, Atkins L, Weinstein HJ (1999) Transplancental transmission of natural-killer-cell lymphoma. N Engl J Med 341(2):85–91
44. Fraser CJ, Hirsch BA, Dayton V, Creer MH, Neglia JP, Wagner JE, Baker KS (2005) First report of donor cell-derived acute leukaemia as a complication of umbilical cord blood transplantation. Blood 106(13):4377–4380, Epub 2005 Aug 23.
45. Nagamura-Inoue T, Kodo H, Takahashi TA, Mugishima H, Tojo A, Asano S (2007) Four cases of donor cell-derived AML following unrelated cord blood transplantation for adult patients: experiences of the Tokyo Cord Blood Bank. Cytotherapy 9(8):727–728
46. Sarugaser ER, Lickorish D, Bash D, Hosseini M, Davies JE (2005) Human Umbilical Perivascular (HUC PV) cells: a source of mesenchymal progenitors. Stem Cells 23:220–229

Chapter 2
Deceased Donors of Tissue

Aurora Navarro

Introduction

Organ and tissue donation are activities that are carried out in many countries. There are some common bases in both activities:

- Both activities are performed by a small number of professionals whose responsibilities are to promote organ and tissue donation at the difficult time for the families facing the death of a relative.
- Donation is an altruist act of the family and anonymity is maintained during the whole process.
- Some organs and tissues may be obtained from living donors but in both cases are not enough to meet demand. In some cases however such as a heart transplant or a cornea is not possible to obtain the graft form a living donor.

On the other hand, big differences exist between organ and tissue donation and transplantation.

- Primarily the purpose of the transplant; organs transplantation is a life saving procedure and tissue transplant is in the majority, life enhancing. This aspect is very important in order to understand the emphasis that is focused for organ donation. Figure 2.1 provides information on donation rates per million population from various countries worldwide. This rate does not meet patient demands. Whilst families understand the need for donation, relatives only accept organ donation, in a more limited way.
- The characteristics of organ and tissue transplants also very different. Organs are given to a patient whilst fresh whilst tissues may be stored for a prolonged period of time and may also be heavily processed. Therefore for organ donation

A. Navarro (✉)
Banc de Sang i Teixits, pg Vall d'Hebron, 119-129 08035, Barcelona, Spain
e-mail: anavarro@bstcat.net

G. Galea (ed.), *Essentials of Tissue Banking*, DOI 10.1007/978-90-481-9142-0_2,

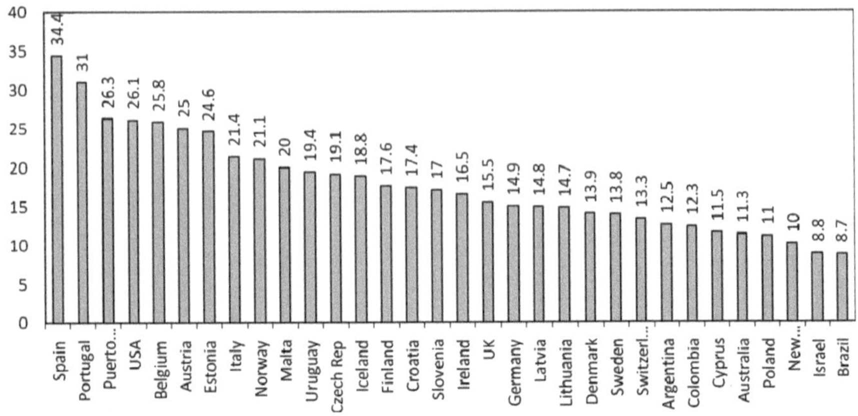

Fig. 2.1 Organ deceased donors 2009. Preliminary data from IRODaT – TPM. Donors with at least one organ recovered and trasplanted. Numbers in pmp (per million population)

the graft and patient survival is often reported back to the donor family while in tissue transplant only tissue outcome is described.

- The number of potential donors is also extremely different between organ and tissue donors. Organ donors can only come from patients dying in the hospital in specific circumstances (e.g. brain death, non heart beating programs etc) and tissue donors has a larger potentiality because patients can die inside or outside of the hospital in varied circumstances. As long as there is compliance with the local legislations and scientific standards related to tissue donor selection tissue can be procured. Figures 2.2, 2.3, 2.4, and 2.5 show the European data on tissue

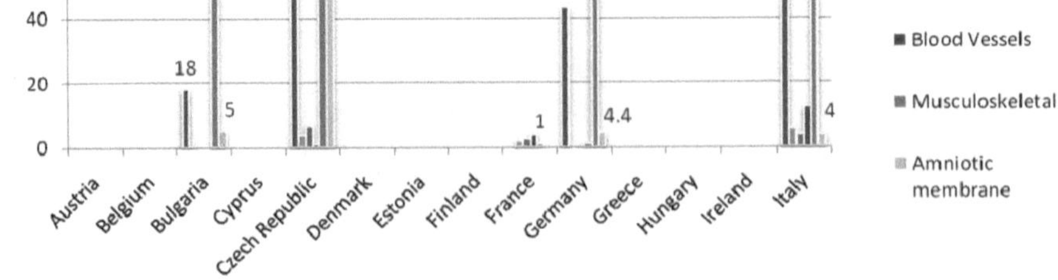

Fig. 2.2 European Union Countries PMP Tissue Donation
Source: www.eurocet.org and Newsletter September 2009, Vol. 14, no. 1.

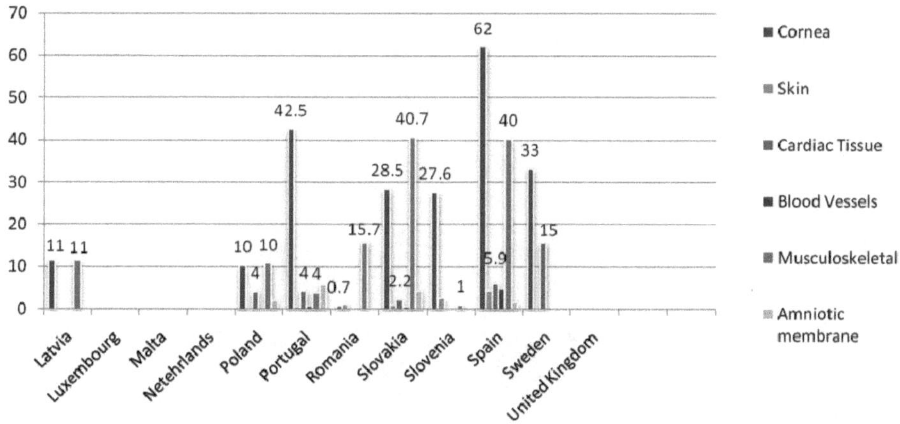

Fig. 2.3 European Union Countries PMP Tissue Donation
Source: www.eurocet.org and Newsletter September 2009, Vol. 14, no. 1.

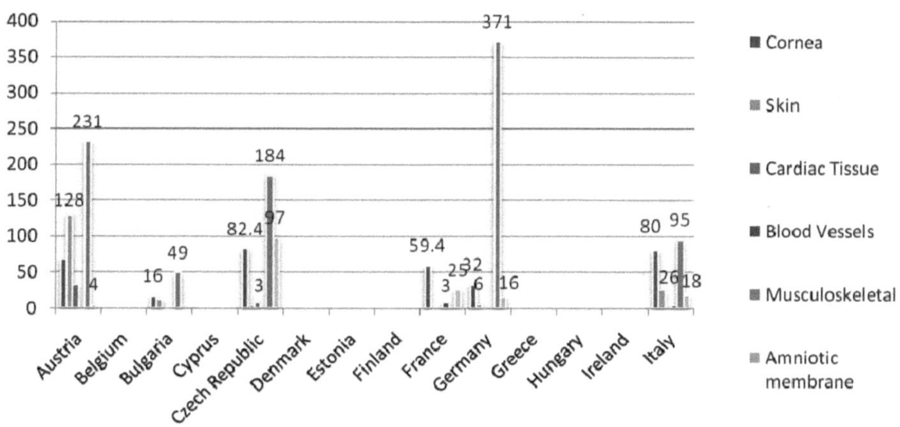

Fig. 2.4 European Union Countries PMP Tissue Transplants
Source: www.eurocet.org and Newsletter September 2009, Vol. 14, no. 1.

collections. They are tissue specific. Because data collection in this area of activity is new, the data is not complete and not all countries have yet reported their activities in tissue banking. It is clear however that there is great variability in tissue collection rates between countries and in the type of tissues procured.

- The time between donation and transplant is also enormously different between organs and tissues. While organs are preserved for a very short time and transplanted in hours, tissues are processed and may be preserved for months, and years. Sometimes the only limitation to preserve a tissue is the validation of the packaging for a longer period of time.

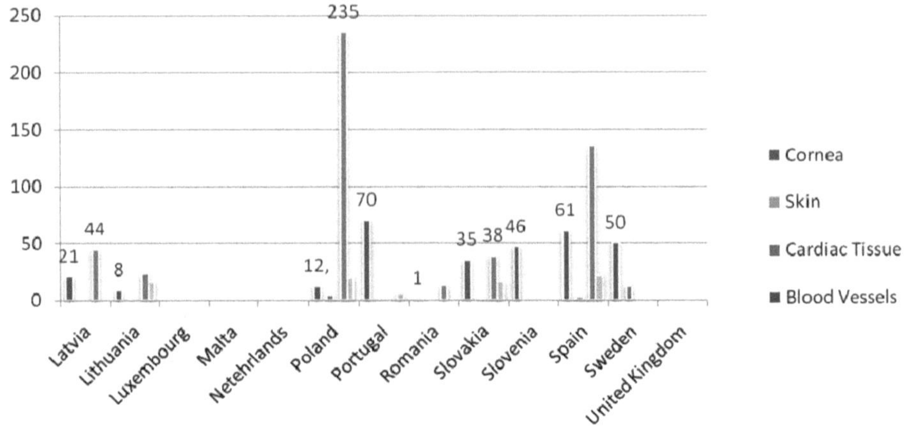

Fig. 2.5 European Union Countries PMP Tissue Transplants
Source: www.eurocet.org and Newsletter September 2009, Vol. 14, no. 1.

Tissue establishments, implement a variety of quality systems to underpin their activities such as ISO systems, Good Manufacturing or Tissue Practices or excellence systems such as EFQM. These quality systems implement directives, laws and scientific standards to ensure the quality of the tissues and minimise the risk of disease transmission through tissue transplantation.

Tissue banks are responsible for maintaining the traceability of the process from donation until transplantation including the contact and solidarity with the family.

Tissue Donor Detection

There are a number of crucial aspects related to tissue donor detection that have to take into account whenever a tissue program is organized:

– Source

The source of the tissue donors will point out where to look for donors.

A hospital tissue donor program will be placed inside the hospital and health care professionals will be part of the process. Usually transplant programs inside the hospital involve health care professional's participation in notifying any potential tissue donor at the moment of death. This program usually starts by obtaining the tissues from all the organ donors and afterwards focuses its efforts to detect potential tissue only donors. Usually this is focussed in areas where deaths occur, usually intensive care units, emergency departments and internal medicine wards. Hospitals' tissue donors are detected soon after death so blood procurement for serologies and tissue recovery is done in a short period of time; usually less than 12 h after death. Also the family interview is usually performed face to face because the relatives are usually still in the hospital when the process starts.

Other strategies include the detection of donors in the anatomo-pathology department. This is based on two important facts: autopsies are performed on a daily basis in most of the hospitals and therefore is a useful source of potential donors. Moreover the pathologist responsible of the autopsy has medical knowledge that will be most welcome during tissue donor selection and physical assessment. The pathologist is in a position to advice the retrieval team and even could participate in tissue recovery.

Finally tissue donors can be detected in the last step after a death occurs, in the funeral home. This system of detecting tissue donors is the most problematic because by the time the potential tissue donor is detected, in many occasions it is too late to start the process of tissue donor selection and retrieval. Also talking to the relatives will be done usually by phone and in some cultures non direct contact between transplant coordinator and the relatives of the potential donors has a higher number of family refusals to tissue donation even it has been proven its efficacy and efficiency in some programs, mostly corneas [1–3].

- Human resources

  Whenever a tissue program starts, it is important to appoint a person responsible for detecting, selecting and following through the whole process of tissue donation. Sometimes different people are involved, administrative, family workers, nurses, doctors. The most important fact is to obtain tissue and organ donors by best utilising their specialization. It is certainly not ideal to perform these activities with personnel whose main remit is in other hospital activities but who are also expected to get involved in tissue donation programmes. Focussing on the professionalism of transplant coordinators and making them the central point has made this activity more efficient not only by increasing the number of tissues retrieved but also by the standard of care to other health care professionals and donor families.

- Structure/dependency/budget

  All the activities developed in the health system need to be clearly structured. Establishing this activity as a core function in a hospital ensures that it is supported by specifically appointed personnel ensures that it is supported by a ring fenced budget.

- Cultural/society

  The culture and the society where we want to implement a program on tissue donation are significant in order to establish the strategies.

A good example of developing a program in one specific country and society is the well known Spanish Model for organ and tissue donation. The Spanish model is not the result of one single act but multiple actions that were established at the beginning of the nineties.

The specific law on brain death diagnosis definition, based on sound ethics and creating a hospital coordinator for organ and tissue donors, who is responsible of detecting and managing donors have been crucial. The direction of this professional is directly from the medical Director of the Hospital. Usually these professionals are intensive care doctors that have partial or complete time dedicated to this activity.

Usually in medium and small hospitals, transplant coordinators have a part time dedication for developing their activities for donation. Other figures depending on the administration are the regional and national transplant coordinators who chair committees to review and analyze organ donation practices.

In order to organize and give national support to the whole process the National Organization for Transplantation (ONT) was established to support the whole system. Since the ONT was established it has been structured by supporting the payment of the centres that generate organs for transplant, a continuous education of the different professionals involved and the support of the mass media which has been cooperating with the National organization with sufficient and correct information to the population from the beginning. Figure 2.6 illustrate the impact of a strategic coordinated plan on the number of family donor interviews and the rate of refusals.

Other systems for developing organ and tissue donation include creating organ procurement organizations (OPO) like in the US where the organizations are placed outside of the hospital and when a death occurs or is imminent there is a requirement to refer the patient to the regional OPO or tissue bank. This referral system is required by federal regulations. The advantage of this system is that the activity is well organized and structured and can meet all the demands. Handling referrals of deceased patients are performed by tissue banking specialist who investigated perform proper risk assessment to screen for tissue donor suitability.

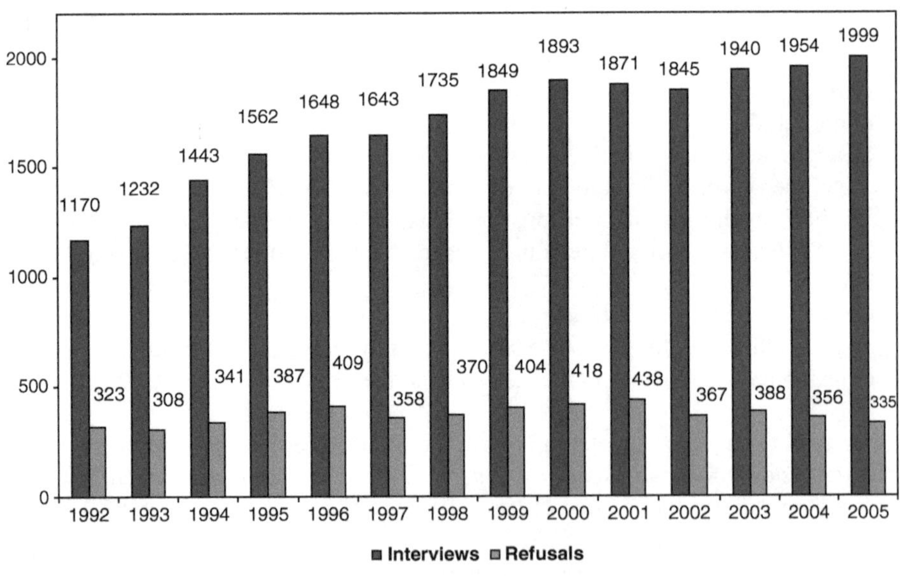

Fig. 2.6 Family refusals and total interviews in Spain 1992–2005. Family refusal rate on 1999 was 27.6 and 16.7% on 2005

– Legal

Laws, specific tissue regulations and European Directives control tissue donation activities from donation until transplantation. Most of these rules establish anonymous, voluntary and unpaid tissue collection contributing to high safety standards and for the protection of human health. Regulation on tissue donation also includes in every country the mechanism to obtain consent for donation through presumed or explicit consent. Presumed consent, mostly accepted in European countries, is based on 'opting out' system offering the possibility to become or not a donor through available registries where individuals can register their desire not to be a donor. In 'opting in' systems individuals have the opportunity to register their intent or consent to donate and in the absence, the default is donation.

Explicit consent or informed consent, mostly in US, UK, Canada and Australia requires the explicit authorization of a donor to donate his/her organs and tissues using donor cards or donor registries that will give his/her intent or consent to donate.

Tissue Donor Selection and Evaluation

As mentioned before, tissue transplantation is of benefit for the recipient but most of the times are not a life saving process. For this reason, the criteria to select a potential tissue donor are more systematic and comprehensive for tissue donors than for organ donors.

This situation is likely to be maintained in the future because organ waiting list for transplantation increases all around the world and some of these patients in waiting list will die while waiting. Tissue waiting lists are less critical because there is the possibility of tissue substitutes in some cases and no mortality in almost all the cases will occur because of a tissue shortage, with the possible exception of skin grafts in severely burnt patients. For example if a cancellous bone is not available, autologous allograft from the iliac crest wedge can be a substitute together with synthetic options. Human heart valves are the first option in some young patients suffering endocarditis, congenital disease but also mechanical valves or could substitute the lack of tissue.

Potential tissue donors are much higher in number than organ donors. Tissue establishments have the advantage that the evaluation of the tissue donor is done during hours, days or even months before clearing the tissue for clinical use. This has positioned tissue establishments in the availability of having clear and specific criteria to determine tissue donor suitability usually before tissue retrieval or in some cases after tissue processing.

Tissue selection criteria have to be followed and established together between the transplant coordinator office working for detecting and selecting the donors together with the tissue establishment that will receive, process and clear the tissues. This agreement should be established from the beginning in order to be efficient during the whole process.

To detect tissue donors is a great effort for the professionals involved because families are shocked facing a very difficult situation and transplant coordinators should focus their efforts to obtaining tissues from tissue donors that the tissue establishment will finally accept.

Nowadays there are a variety of different tissue selection criteria- documents generated in individual countries, European directives and standards from different scientific organizations. All of them establish the basis to select a good quality tissue which at the same time that minimizes the risk of disease transmission.

These documents have different approaches to donor selection depending on their legal standing. Scientific organizations standards are recommendations and guidelines very useful for practitioners in tissue banking field. Laws, Directives and states regulations are legally binding and have to be enforced by the tissue establishment in the influence area.

Whilst historically, countries have established their own requirements for tissue donor selection, retrieval and banking in their territory, The European Parliament and the Council of the European Union have produced directives ensuring that the tissue banking activities are performed to harmonised minimum standards. Today these activities in Europe are covered by the requirements laid out in European Directive 2004/23/EC, of the European Parliament and the Council setting standards of quality and safety for the donation, procurement, testing, processing, preservation, storage and distribution of human tissues and cells and more recently [4].

In order to implement the mother Directive, Directives 2006/17/EC regarding certain technical requirements for the donation, procurement and testing of human tissues and cells and 2006/86/EC was published. Finally 2006/86/EC implements certain technical requirements concerning traceability, notification of serious adverse reactions and events, and certain technical requirements for the coding, processing, preservation, storage and distribution of human tissues and cells [5, 6].

These documents, are legally binding and are intended to apply to all the processes that take place in a tissue establishment, consisting of donor selection, procurement, processing, storage and distribution of tissues and cells for human application. In this chapter we will focus on all the aspects related to complete donor selection criteria in order to have a safe tissue to be transplanted.

In order to perform complete tissue donor selection criteria all these activities must be conducted:

- Review donor's medical history
    - General donor exclusion criteria
    - Tissue absolute contraindications
- Family consent details and donor's social history
- Donor's physical examination
- Blood sample collection
- Biopsy/autopsy results

Review Donor's Medical History

A global acceptance of the potential tissue donor has to be done before facing the decision of the specific target tissues to be retrieved.

The global analysis is not only a checking list of the medical history of the donors compared to the exclusion criteria list. Donors often have a multiplicity of diseases that have to be interpreted by an expert.

Different data has to be obtained to have a complete medical history: hospital medical chart, family doctor report and recent analysis. For this reason it is recommended that an expert on tissue donor selection takes the final decision, usually the Medical Director of the tissue establishment.

Whenever a decision is being taken it has to be done under objective data of the information and a risk analysis of the decision should be done. The risk assessment should be is evaluated on the potential severity of risk of harming the recipient through a tissue transplant and the relative prevalence of such an occurrence.

General Donor Exclusion Criteria

– Malignancies are excluded except in situ cancer of the cervix and basal cell carcinoma and some primary brain tumours which are accepted for tissue donation. For cornea donation also solid tumours are accepted if there is no metastasis to the anterior chamber of the eye or the primary tumour has tropism for producing so. Also retinoblastoma and haematology cancer should exclude cornea donation.

  This criterion is in most of the countries' guidelines and actually the EU Directive excludes cancer for the rest of tissues in Directives 2006/17/EC. Nowadays some tissue establishments are considering accepting tissue donors with cured cancers. In the AATB standards for tissue banking [7] there is a section which gives guidance that "donors with current or prior malignancy shall be evaluated by the Medical Director or licensed physician designee for suitability in accordance with the tissue bank's SOPs. The evaluation shall include the type of malignancy, clinical course, and the treatment prior to acceptance of a donor. The evaluation and reasons for acceptance shall be documented in the donor's record". These criteria adopted by the American standards could not be adopted by any European countries covered by the EU Directive.

The following brain tumors that are accepted for tissue donation are:

- Meningioma
- Pineocytoma
- Craniopharingioma
- Adenoma
- Teratoma

- Unknown cause of death or diseases of unknown aetiology will exclude tissue donation. Usually if the cause of death is unknown an autopsy will be performed. In these cases if the cause of death is very likely to be found, tissue retrieval can be performed waiting for the final result of the autopsy. In other complicated cases where is not clear that the result will be resolved the medical director can decide not to proceed tissue retrieval
- Uncontrolled infection such as septicaemia, systemic mycosis and viral diseases will exclude tissue donation. The analysis of an uncontrolled infection has to be done in conjunction with the treating doctor in order to analyse the risk of retrieving an infected tissue. Even though cultures of the tissues are taken during retrieval and through processing, it is recommended not to retrieve tissues from uncontrolled infected donors if no validated inactivation process is performed. Some tissue banks complement the information of potential infection with blood cultures of the donors, although the utility of this is still debatable [8–12]. When there is a suspicion, history, risk factor or evidence of the presence of HIV, HCV or HTLVI-II the donor should be excluded. HTLVI/II geographical area distribution is Japan, northern part of South America, Caribbean, southeast USA, some areas of Central Africa, Middle East and India. Mainly in USA and Europe patients with HTLVI/II are drug addicts and their partners.
- Neurodegenerative diseases of unknown aetiology such as lateral Amniotrofic Sclerosis, Parkinson's disease and Multiple Sclerosis will exclude tissue donation. Neurodegenerative disease have unknown physiopathology in most cases and usually are related to genetic components, ambient factors or even infectious origin such as viruses.
- Family history of CJD and vCJD, dementia of unknown causes, and previous treatment of growth hormone and infertility before 1985 as well as the receipt of a dura mater, cornea or sclera should exclude the donor from tissue donation because of the risk of prion transmission.
- Autoimmune diseases. If the diseases involves different systems which may affect the quality of the tissue, the tissue should not be donated. It is also important to note that if the patient is under immunosuppressant treatment the results of the serologies for the mandatory markers may be invalid.
- Xenotransplant (tissues containing living animal cells) recipients should not donate tissues

Tissue Absolute Contraindication and Age Criteria

Once the donor has been evaluated overall a close tissue specific selection criteria has to be performed.

Tissue banks establish age criteria to collect viable tissues with a high rate of probability for being transplanted. For this reason age criteria are not standardized among different tissue banks because it depends of the need and waiting list of tissues and the characteristics of the target potential tissue donor.

In order to represent some data for specific tissue exclusion criteria see Table 2.1. This is by no means exhaustive but the following is illustrative of the types of conditions that need to be assessed.

Heart Valves

Pulmonary Valve: <65 years old
Aortic Valve: <50 years old.
Mitral valve: <50 years old

Heart Valve donor should meet the following criteria:

- No history of bacterial endocarditis, coronary artery bypass grafting, rheumatic fever, semilunar valvular disease or cardiomyopathy of unknown aetiology or viral.
- Evaluation of risk of Chagas' disease
- Evaluate every case of previous cardiac surgery, closed chest massage, penetrating cardiac injury
- Marfan Syndrome
- Evaluate history of mitral valve diseases, including valve prolepses

Vascular

Arteries: <50 years
Veins: <60 years

Vascular donors should not be accepted if:

Arteries: history of arteriosclerosis or trauma in site. If more than three medical antecedents related to arteriosclerosis (high blood pressure, diabetes, smokers, obesity and dislipemia) are found, arteries should be contraindicated.

There should not be history of vein stripping, varicose veins or legs ulcers.

Skin

Skin <75 years old

Physical assessment shall include documentation of findings that may affect the quality or quantity of skin to be retrieved.

Musculoskeletal

Cancellous bone: no age limit
Osteoarticular allografts and tendons: <55 years old

The following medical criteria should be contraindication:

- History of sarcoidosis, systemic lupus erytrematosus, rheumatoid arthritis or any clinical metabolic bone diseases.
- Chronic steroid treatment is linked to a higher risk of osteoporosis.
- Diagnosed osteoporosis
- Toxicity from some substances such as cyanide, lead or mercury.
- Irradiation should be considered

Eye Tissues

Corneas: no age limit, though most tissue establishments establish age criteria to maintain an effective viable rate for transplantation.
Sclera: <60 years old.

The following findings should contraindicate corneal tissue donation:

Antecedents of keratitis, anterior chamber melanoma
Anterior segment pathologies: corneal scars, dystrophies, keratoconus, ectasias and leucomas.
Retinoblastoma, anterior chamber melanoma and any tumour that can have metastasis in the eye.

Family Consent Details and Donor's Social History

Consent has to be performed by the legally appropriate next of kin. Information about tissue donation should be comprehensive. The interview with the relatives shall take place in surroundings where the relative is free to ask questions about the donation process and the transplant coordinator is responsible for giving precise information in accordance with the country's or state's law.

After obtaining consent, the family interview is the starting place for information related to the potential donor in the following items: verifying the medical history of the donor, confirming the source of different findings, evaluate the behavioural risk factors, assess the sexual risk and information about travelling. Also it is necessary to inform the family about the need to state that blood samples from the donor will be tested for certain transmissible diseases.

Table 2.1

Heart valves	Arteries	Skin	Musculoskeletal	Cornea
<60–65	<50	<75	No age limit cancellous tendons < 55	No age limit
Semilunar valvular disease Obstructive hypertrophic myocardiopathy Endocarditis Marfan disease Coronary artery by-pass Rheumatic fever	No arterioesclerosis Trauma HIgh blood pressure Diabetes mellitus Smokers Obesity	Connective tissue pathology Nevus Lesions Hipertricosis Ingestion or exposure to toxic substances	Tissue infection Tissue toxicity Long term steroids treatment Osteoporosis Prosthesis Rheumathoid arthritis Systemic lupus eritematosis Poliarteritis nodosa Sarcoïdosis Clinically significant metabolic bone disease	Retinoblastoma and melanoma anterior chamber Quratitis antecedents HSV Ocular metastasis Ulcers Active infections anterior chamber Scars, pterigion

To perform all these investigation sometimes is necessary to find out the best interlocutor that sometimes is not the same person who's giving consent.

A questionnaire to ask the relative is recommended in order to look for this information and it is suggested to be presented once the family has already accepted that donation will go ahead. In this situation the family understands easily that all the questions that will come forward will secure the final transplant of the tissues.

The different questions in the questionnaire should cover all the circumstances considered to provide a high incidence of risk for disease transmission. Not all the criteria for preventing transmission of relevant communicable disease agents through transplantation of tissues and cells are the same in different countries and therefore the standards will differ depending on the local epidemiologies.

These criteria should be based on the experience of the target society. Some behaviour/history exclusion criteria will exclude donors permanently, or for at least 5 years and others for a minimum of 12 months.

Exclude if in the previous 5 years/ever

- men who had sex with another men.
- Persons who has injected drugs
- Persons who have had sex in exchange for money or drugs

In the preceding 12 months:

- Person who have had sex with any person from the 3 items described above
- Persons who have been exposed to known or suspected HIV, HBV, HCV infected blood through percutaneous inoculation or through contact with an open wound
- Children born to mothers known to be HIV infected or risk for HIV infection, who are 18 months of age or less and/or breast feeding in the preceding 12 months.
- Person who have lived with another person having viral hepatitis
- Persons who had or have been treated for syphilis or gonorrhea
- Persons who undergone tattooing, acupuncture, ear or body piercing done with shared instruments
- Current inmates of correctional systems and individuals who have been incarcerated for more than 72 h.

The time limit to accept potential donors under any of the listed categories is under constant debate. With the introduction of highly sensitive molecular testing, window periods have shrunk to a period of days. The question is often asked - why should potential donors be deferred for such long periods, therefore? The reasons are complex but include the possible transmission of viruses that may be transmitted that are currently not tested for

Finally questions about travelling or exposure to areas for specific infections such as West Nile Virus, malaria and Chagas' disease should be requested. Specific test for these diseases should be performed if the donor was in these areas for a period of time.

Donor's Physical Examination

This is another step to evaluate the medical and social history or even find out information not reported previously.

Tissue donor physical exam shall always be performed by a professional whose responsibility is to detect through this evaluation any evidence of risk behaviour or any sign of infection or viral disease. Some of the findings listed below, if found, makes it necessary to ask the family about the time and circumstances of the sign (e.g. piercing, tattoo). If there are other critical signs found, the donor shall be rejected.

The physical examination is a systematic activity that has to be performed in any circumstance and always in the same way; from head to toe or the opposite way but always in a systematic way and knowing exactly what you are looking for. There are 6 signs that should always be specifically looked for, as a minimum.

Sign 1. Physical evidence for risk of sexual transmitted diseases: syphilis, ulcers, herpes simples, chancroid, and peri-anal lesions.
Sign 2. Physical evidence of non-medical percutaneous drug abuse, acupuncture or tattoos including the presence of ear or body piercing.
Sign 3. Significant enlarged lymph nodes
Sign 4. Oral thrush, blue spots or purple ones characteristics of Kaposi's sarcoma.
Sign 5. Generalized rash that could point to sepsis or unexplained jaundice.
Sign 6. Necrotic lesions after vaccination

Donor's physical examination shall be recorded in the donor's history and corroborated by the retrieval team.

Blood Sample Collection

Collecting samples from deceased donors who are not organ donors raise some difficulties in relation to the quality of the blood procured.

The quality of the blood to carry out serologies is directly related to two aspects: the time of blood collection and plasma dilution.

The collection of blood samples shall be before death or as soon as possible after death – always under 24 h after asystole. Taking samples after 24 h of death will produce a higher number of false positive and samples collected as soon as possible after death will produce a least number of false positive results [13–15].

If the donor has received massive blood transfusions or colloids in the previous 48 h or more than 2,000 mls of crystalloids in the previous hours or any combination of both, prior to asystole, the results of the serolgies can show false negative results because of the plasma dilution. In these cases it is necessary to find a sample prior to the administration of blood or fluids.

If there is no blood sample available it is necessary to apply an algorithm to show if the donor's plasma is diluted more than 50%. See the algorithm algorithm below.

48 hours previous blood collection

A. Total Blood transfusion=. . . . . .ml
B. Total Colloid transfusions=. .ml

1 hour previous blood collection
C. Total Cristalloid=. . . . .ml

Donor data
PV: Plasmatic volume=weight/0,025. . . . .ml
BV: Blood volume= weight/0,015. . . . .ml

If B + C are bigger than PV or A+B+C are bigger than BV contraindicate the sample

The serology test must be performed in an accredited laboratory and using validated kits. There are some validated kits for hepatitis B surface antigen for samples from cadaveric donors.

After testing, a sample is necessary to archive the serum. This archive is required to ensure a retrospective testing at any time if there is any suspicion of disease transmission through tissue transplantation or because a new disease is necessary to be tested in all the stored tissues. The storage of the samples has to last for at least ten years after the last tissue is transplanted.

The serologies required for tissues coming from cadaveric donors usually include:

- hepatitis B surface antigen (HBsAg)
- antibody to hepatitis B core antigen (anti-HBc)
- antibodies to hepatitis C virus (anti-HCV)
- antibodies to human immunodeficiency virus type 1 and 2 (anti-HIV-1 and anti-HIV-2)
- syphilis
- antibodies to human T-lymphotropic virus type I and II (anti HTLV-I and anti HTLV-II

In order to diminish the window period NAT testing is recommended for HIV and HCV.

If any of the tests above are positive, the tissues should be discarded excepting anti-HBc and syphilis where, additional tests can be performed.

AATB standards do not accept anti-HBc positive and recommend to discard all the tissues with anti-HBc positive result appears.

In other standards and regulations, algorithms are established. When anti HB-c is positive, additional test must be performed such as anti-HBs Antibodies and NAT testing. In the case of a positive syphilis result, an appropriate algorithm will exclude an active infection for Treponema Pallidum.

Biopsy/Autopsy Results

Postmortem evaluation and report (when available) need to be included with the tissue donor's documents complementing the information gathered during the whole donation process. When a simultaneous organ retrieval occurs a systematic inspection is done during organ retrieval but when it is an only a tissue retrieval it is recommended that an autopsy is carried out to identify any latent pathologies.

Conclusion

Obtaining tissues from cadaveric donors is a complex process involving many healthcare professionals. The success of such a programme depends on a number of factors but is crucially dependent on an overarching national strategy. Numerous tissues can be obtained from these donors and systems are in place to ensure their quality and safety. All these activities are now supported by quality systems and regulations.

References

1. Geissler A, Gerbeaux PR, Maitrejean C, Durand-Gasselin J (2005) Cornea donation: evaluation of a training session to obtain consent by telephone. Transplant Proc Dec;37(10): 4634–4636
2. Rodríguez-Villar C, Ruiz-Jaramillo MC, Paredes D, Ruiz A, Vilardell J, Manyalich M (2007) Telephone consent in tissue donation: effectiveness and efficiency in postmortem tissue generation. Transplant Proc Sep;39(7):2072–2075
3. Gain P, Thuret G, Chiquet C, Pugniet JL, Rizzi P, Tchaplyguine F, Acquart S, Le Petit JC, Maugery J (2002) Cornea donation consent by telephone. J Fr Ophtalmol 25(6):577–583, Jun 2002
4. Directive 2004/23/EC of the European Parliament and of the Council of 31 March 2004 on setting standards of quality and safety for the donation, procurement, testing, processing, preservation, storage and distribution of human tissues and cells (7th April 2004)
5. Commission Directive 2006/17/EC implementing Directive 2004/23/EC of the European Parliament and of the Council as regards certain technical requirements for the donation, procurement and testing of human tissues and cells (8 February 2006)
6. Commission Directive 2006/86/EC implementing Directive 2004/23/EC of the European Parliament and of the Council as regards traceability requirements, notification of serious adverse reactions and events and certain technical requirements for the coding, processing, preservation, storage and distribution of human tissues and cells (24 Oct 2006)
7. American Association of tissue banks. Standards for tissue banking, 11th edn, 2006. Section D. Acquisition of tissue consent, donor screening, and tissue retrieval. D4.340 Malignancies

8. Saegeman V, Verhaegen J, Lismont D, Verduyckt B, De Rijdt T, Ectors N (2009) Influence of postmortem time on the outcome of blood cultures among cadaveric tissue donors. Eur J Clin Microbiol Infect Dis. Feb;28(2):161–168. Epub 2008 Aug 16
9. Malinin TI, Buck BE, Temple HT, Martinez OV, Fox WP (2003) Incidence of clostridial contamination in donors' musculoskeletal tissue. J Bone Joint Surg Br Sep;85(7):1051–1054
10. Vehmeyer S, Wolkenfelt J, Deijkers R, Petit P, Brand R, Bloem R (2002) Bacterial contamination in postmortem bone donors. Acta Orthop Scand Dec;73(6):678–683
11. Vehmeyer SB, Bloem RM, Petit PL (2001) Microbiological screening of post-mortem bone donors – two case reports. J Hosp Infect Mar;47(3):193–197
12. Martinez OV, Malinin TI, Valla PH, Flores A (1985) Postmortem bacteriology of cadaver tissue donors: an evaluation of blood cultures as an index of tissue sterility. Diagn Microbiol Infect Dis May;3(3):193–200
13. Padley D, Ferguson M, Warwick RM, Womack C, Lucas SB, Saldanha J (2005) Challenges in the testing of non-heart-beating cadavers for viral markers: implications for the safety of tissue donors. Cell Tissue Bank;6(3):171–179
14. Heim A, Wagner D, Rothämel T, Hartmann U, Flik J, Verhagen W (1999) Evaluation of serological screening of cadaveric sera for donor selection for cornea transplantation. J Med Virol Jul;58(3):291–295
15. Aswad S, Khan NS, Comanor L, Chinchilla C, Corado L, Mone T, Mendez R, Mendez R (2005) Role of nucleic acid testing in cadaver organ donor screening: detection of hepatitis C virus RNA in seropositive and seronegative donors. J Viral Hepat Nov;12(6):627–634

Chapter 3
Banking of Cord Blood

Eliane Gluckman

Introduction

Hematopoietic stem cell transplantation (HSCT) can be curative in a large variety of selected malignant and non malignant diseases. Umbilical cord blood transplantation (UCBT) has extended the availability of allogeneic hematopoietic stem cell transplantation (HSCT) to patients who would not otherwise be eligible for this curative approach. The first successful UCBT from an HLA-identical sibling in a child with severe Fanconi's anemia was reported by Gluckman et al. in 1989 [1]. This first success opened the way to a new field in the domain of allogeneic HSCT as it showed that: (1) a single umbilical cord blood contained enough hematopoietic stem cells to reconstitute definitely the host lympho-hematopoietic compartment; (2) an umbilical cord blood unit could be collected at birth without any harm to the new-born infant, and (3) umbilical cord blood hematopoietic stem cells could be cryopreserved and transplanted in a myeloablated host after thawing without losing their repopulating capacity. Since, our knowledge on the biological characteristics of umbilical cord blood cells has increased, emphasizing the advantages of using umbilical cord blood stem cells for transplant. Simultaneously, umbilical cord blood banks (CBB) have been established for related or unrelated UCBT with more than 500,000 units available and more than 20,000 umbilical cord blood transplants performed in children and in adults with malignant and non malignant diseases.

Umbilical cord blood has many theoretical advantages due to the immaturity of newborn cells. Hematopoietic progenitors from umbilical cord blood are enriched in primitive stem cells, which are able to produce in vivo long-term repopulating stem cells [2]. Compared to adult cells, umbilical cord blood hematopoietic stem cells grow larger colonies, have different growth factor requirements, and are able to expand upon long term culture in vitro; The properties of umbilical cord blood cells should compensate the relatively low number of cells contained in a single umbilical cord blood and, through rapid expansion, reconstitute myeloablated

E. Gluckman (✉)
Eurocord, Hematology Department, APHP University Paris VII, Paris, France
e-mail: eliane.gluckman@sls.aphp.fr

G. Galea (ed.), *Essentials of Tissue Banking*, DOI 10.1007/978-90-481-9142-0_3,

patients. Despite the capacity for cord blood cell expansion, clinical results showed that hematopoietic recovery was delayed after cord blood transplants; engraftment was associated with the number of nucleated and CD34+ cells infused and the number of HLA differences [2]. However, a recent study showed a higher frequency of early and committed hematopoietic progenitors in long term hematopoietic reconstitution in children receiving a cord blood compared to a bone marrow allogeneic transplant [3].

The second advantage of UCBT relates to the immaturity of the immune system at birth. This property should decrease the lymphocytes alloreactive potential and, consequently, reduce the incidence and severity of graft versus host disease (GVHD) after an HLA matched or mismatched transplant. Cord blood lymphocytes are naive and immature, most functions are inducible through in-vitro or in-vivo activation; early NK and T cell cytotoxicity is impaired but secondary activation can occur. Therefore, one can speculate that, despite the reduction of GVHD, an anti leukemic effect – called graft versus leukemia (GVL) – can still be observed. As acute GVHD is an early event after allogeneic bone marrow transplant (BMT) and is in part triggered by cytokine release, it is reasonable to postulate that UCBT induces less frequent and less severe acute and chronic GVHD than adult HSCT which contain a higher number of activated T cells [4]. These properties should lead to less stringent criteria for HLA donor-recipient selection.

In comparison with other sources of allogeneic HSCT, UCB offers substantial logistic and clinical advantages such as (1) significantly faster availability of banked cryopreserved UCB units, with patients receiving UCB transplantation in a median

Table 3.1 Searching and identifying an alternative stem cell donor. Main criteria to be considered

	UBMT	UCBT	Haplo-HSCT
Information on A + B + DRB1 typing (%)	16–56	~80	100
Median search time (months)	3–6	<1	Immediate
Donors identified but not available (%)	20–30	~1	None
Rare haplotypes represented (%)	2–10	20	Not applicable
Main limiting factor to graft acquisition	HLA identity	Cell dose	Poor mobilization
Ease of rearranging date of cell infusion	Difficult	Easy	Easy
Potential for immunotherapy	Yes	No	Yes (limited)
Potential for viral transmission to recipient	Yes	Yes	Yes
Potential for congenital disease transmission	No	Yes	No
Risk for the donor	Low	No	Low
Main problems to be overcome	GVHD	Engraftment	Immune deficiency relapse

of 25–36 days earlier than those receiving BM [2] (2) extension of the donor pool due to tolerance of 1–2 HLA mismatches out of 6 (higher HLA mismatched is associated with lower probability of engraftment) (3) lower incidence and severity of acute graft-versus-host disease (GVHD); (4) lower risk of transmitting infections by latent viruses, such as cytomegalovirus (CMV) and Epstein-Barr virus (EBV); (5) lack of donor attrition; (6) lack of risk to the donor and finally (7) higher frequency of rare haplotypes compared to bone marrow registries, since it is easier to target ethnic minorities [5, 6]. The disadvantages of UCBT are (1) the low number of hematopoietic progenitor cells and HSCs in UCB compared with BM or mobilized peripheral blood stem cells (PBSC) that translates in increased risk of graft failure and delayed hematopoietic engraftment and (2) the impossibility of using donor lymphocyte transfusion for immunotherapy. The main differences between the advantages and disadvantages of using bone marrow and cord blood are listed in Table 3.1.

Cord Blood Collection and Banking

Cord Blood Banks Development

The progress in the field of umbilical cord blood transplantation is parallel with the huge interest in establishing and developing cord blood banks worldwide. Today, more than 500,000 cord blood grafts are available in more than 100 cord blood banks. These banks play an important role in the process of cord blood transplantation. The Netcord group was created in 1998 to establish good practices in umbilical cord blood storage, facilitate donor search, improve the quality of the grafts, standardize excellence criteria on an international scale and importantly establish procedures for bank accreditation in collaboration with FACT (Foundation on Accreditation in Cell Therapy) [7–9]. The inventory of Netcord, the cooperative network of large experienced UCB banks, currently has more than 200,000 cryopreserved UCB units ready for clinical use for unrelated recipients and more than 8,624 grafts shipped (Table 3.2). Recently, the National Marrow Donor Program (NMDP) has established a similar cord blood bank network in the USA with the financial support of the American Congress. Collaborations between Netcord-Eurocord and NMDP have been established with the goal to provide the most appropriate and high quality cord blood unit for a specific patient.

Other types of cord blood banks have been established such as sibling donor cord blood banking or autologous (or commercial family CB banking) where there is no existing family indication for HSCT. There are 2 types of CBB, public and private according to their economic interest and financial support and 3 types of CBB according to the type of donation and use, unrelated, sibling donor or autologous CBB. Unrelated donor transplantation programs employ public banks as their source of donor cord blood units (CBU). These CBUs are donated on a volunteer basis by women delivering healthy babies at term. Private Banks, which are

Table 3.2 Netcord inventory

NETCORD Inventory and Use September 2008

CB Bank	Inventory	Released for Transplant	Children	Adults
NETCORD/FACT accredited				
Sydney + Melbourne	16044	476	239	237
Barcelona	9831	436	186	240
Düsseldorf	14344	540	264	251
Durham	19056	898		
France	6585	913	283	630
Helsinki	2930	19	10	9
Houston	5555	83	40	43
Liege	1994	114	47	67
London	9907	226	125	101
Milan	7134	370	194	176
New York	43385	2730	1724	1006
Pavia	2200	89	36	53
Sum	138965	6893	3147	2813
not NETCORD/FACT accredited				
Athens	1036	6	5	1
Brisbane	3715	36	17	19
Firenze	1085	63	31	22
Gauting	2374	46	18	28
Leiden	3642	80	35	45
Leuven	8308	120	63	57
Louvain	1862	89	33	56
Brussels	1313	26	9	17
Málaga	11829	92	32	60
Mannheim	1661	29	17	12
Mexico City	1237	101	68	33
Padova	1379	47	18	29
Pescara	375	3	3	0
Prague	2947	22	10	12
Roma Lazio	1225	48	26	22
Santiago de Compostela	4997	49	27	21
Seoul	5755	23	10	13
Tel Hashomer	1463	18	10	8
Tokyo	5324	833	226	607
Sum	61527	1731	658	1062
TOTAL	200492	8624	3805	3875

Official NETCORD Collaborators: thermogenesis biosafe AIR PRODUCTS

H. Gressmann, NETCORD Virtual Office, Duesseldorf

for-profit entities, store "directed donations" collected by obstetricians from babies born into families who intend to use the cord blood for the baby from whom it came (autologous donation) or for another family member in need of future transplantation therapy. Sibling donor cord blood banks (SDCBB) programs have been established in a public or private model for families in whom there is an indication of hematopoietic stem cell transplant. The differences between these 3 types of banks are shown in Table 3.3.

Unrelated Cord Blood Banks

The number of unrelated cord blood banks and, in consequence, the number of available cord blood units for unrelated use is increasing worldwide. We estimate that more than 500,000 cord blood units are available for transplantation in more than 100 cord blood banks in many countries (www.bmdw.org, assessed on November 2008). Summary of cord blood banks activity is shown in www.wmda.org and summarized in Figs. 3.1 and 3.2. The price of a cord blood unit varies between 15,000 and 35,000 Euros. Currently, there is an increasing number of international exchange cord blood units. For example in France, from January to October 1st 2008, out of an inventory of 6,586 units collected in 3 cord blood banks, 290 units

Table 3.3 Comparison of various banks models

	Autologous	Unrelated	Sibling
Operating model:			
Financial status	For-profit	Non-profit	Non-profit
Donation	Business transaction	Voluntary act	Motivated act
Donor's status	Payer	Neutral	Non-fiduciary benefit
Product ownership	Contingent upon payment?	Public	Family
Operating location	Remote sites	Few specified sites	Remote sites
Potential size/market	Very large	Large	Small
Safety and QA:			
Suitability	All who can pay	Medical and lab deferrals	Few absolute deferrals
ID testing	Unpublished	Full blood donor ±NAT[a]	Donor re-tested ±NAT
Genetic disease testing	Unpublished	Yes, expanded testing under consideration	Established, often specific to family's mutation
Deliberate maneuvers to facilitate CB collection volume	No	No	Not excluded
Donor-recipient linkage	Defining property	Under debate	Intrinsic
Trained collectors	No	Yes	Limited training at the time of collection
Quality assurance	Unpublished	Established	Under development using blood center model
Disposition of units:			
Likelihood of complete histocompatibility	100%	Low	~25%
Likelihood of use for transplant	Low to nil	Moderate	Relatively high
Research potential	Low to nil	High	Moderate to high
Crossed-over for public use	No	n/a	Under discussion

[a] Nested Antigen Testing

were released 115 for French patients and 59 for foreigners. During the same period 175 (65%) were provided by foreign banks (Source Agence de la Biomédecine).

National regulatory agencies and transplant centers are aware of the need of international standards for cord blood collection, processing, testing, banking, selection and release. In 2006, the Netcord-FACT published the third edition of the International standards for cord blood. Founded in 1998, Netcord is the international cord blood banking arm of Eurocord. The mission of Netcord is to promote high quality cord blood banking and clinical use of umbilical cord blood for allogeneic stem cell transplantation. Approximately 20 CBB, mostly all European CBB

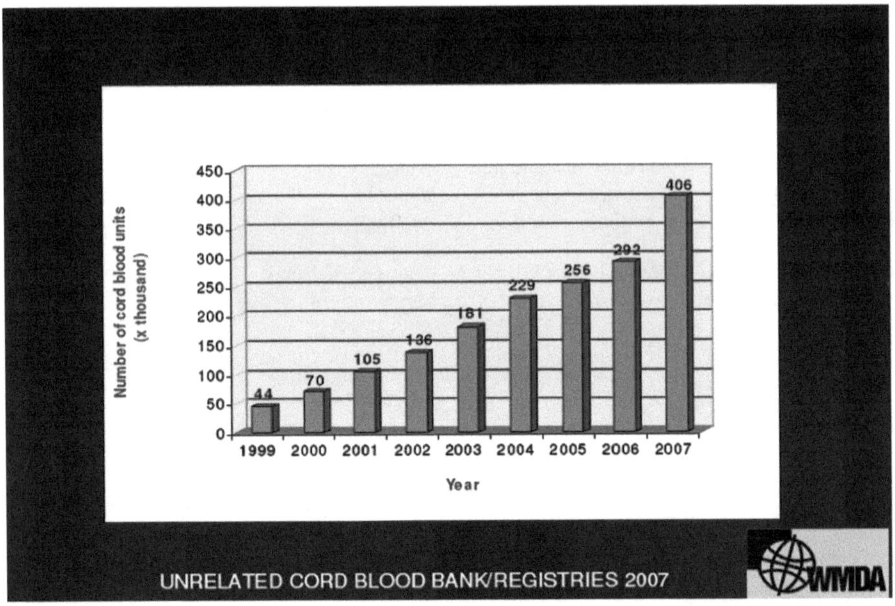

Fig. 3.1 Number of cord blood units available for unrelated transplantation on 31-12-2007

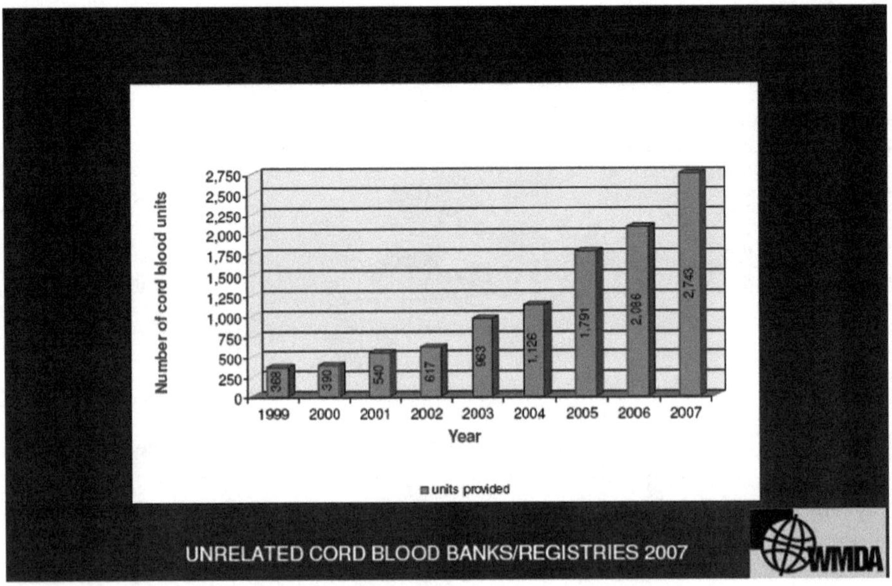

Fig. 3.2 Total number of cord blood units provided

are Netcord members accounting for almost 50% of worldwide available units. To be an active Netcord member, among other criteria, a Netcord-FACT accreditation is required. Some of Netcord CBB has been already accredited and others are going through the process. The major objective of these standards is to promote quality medical and laboratory practices throughout all phases of cord blood banking with the goal to achieve consistent production of high quality placental and umbilical cord blood units for transplantation. These standards cover (1) collection of cord blood cells, regardless of the methodology or site of collection [10, 11]; (2) screening, testing, and eligibility determination of the maternal and infant donor according to Applicable law; (3) all phases of processing and storage, including quarantine, testing, and characterization of the unit; (4) making the CB unit available for transplantation, either directly or through listing with a search registry; (5) the search process for selection of specific cord blood units; and (6) all transport or shipment of cord blood units, whether fresh or cryopreserved [12, 13]. To be compliant with Standards CBB must use validated methods, supplies, reagents, and equipment; maintain a comprehensive; properly documented Quality management Program; and track the clinical outcomes of patients who receive cord blood units from that bank. The accreditation process includes submission of written documents and on site inspection of collection, processing, and storage facilities. Netcord-FACT accredited CBB are reinspected routinely every 3 years.

All the practical aspects of cord blood banking, such as maternal informed consent, collection techniques, labelling and identification, infectious disease and genetic disease testing, HLA typing, methodology of cell processing, cryopreservation, transportation and release have been extensively published . All these aspects are detailed in the last version of the Netcord-FACT Standards (www.factwebsite.org).

As the number of cord blood units is increasing it appears that it is necessary to improve the quality of the units for cost efficient management of the banks. The optimal number of cord blood units is not really known but should approach 9 for 100,000 inhabitants. Most banks prefer to collect only the largest units of more than 70 mL in order to obtain at least 3×10^7 nucleated cells/kg. The effect of increasing the inventory from 50,000 to 300,000 for finding a matched cord blood with a minimum cell dose of 2.3×10^7 per kg results of an increment of chance to find a donor of 19% for children and 10% for adults [14–16]. This is because children have a lower weight increasing their chance to have a donor with the appropriate number of cells.

Sibling Donor Cord Blood Banks

Sibling donor cord blood banks (SDCBB) programs have been established for families in whom there is an indication of hematopoietic stem cell transplant. This, because there is a 25% probability that a new sibling will be human leukocyte antigen (HLA)-identical to the existing full sibling, a family and their child's physician often agree that it is prudent to collect and cryopreserve the new sibling's CB in the event that a clinical indication for CB transplantation develops in the affected child. Currently most of the cord blood units coming from a sibling cord blood donor is collected and frozen in hospitals where the family recipient is being treated. In order to unify and harmonise collection and cord blood unit processing following regulatory procedures sibling donor cord blood banks have been established. Because these relatively few families give birth at a variety of smaller hospitals, it is a considerable logistical challenge to provide access to specialized high quality SDCB banking services for them. An example of this type of national banking program is the SDCBB at Children's Hospital Oakland in USA [17, 18]. Many of the operational procedures and medical policies have been developed for national comprehensive SDCB banking program.

The scope of need for SDCB banking services, though far smaller than unrelated donor CB programs, is considerable. Approximately 5,000 children in the US age 0–14 are newly diagnosed each year with a malignant process and approximately 40% of these represent hematopoietic malignancies [19, 20]. Enrolment in Oakland's SDCB program is consistent with this epidemiology; nearly half the enrolments are by families caring for a child with a hematopoietic malignancy. Related cord blood banking is particularly cost effective in families when a child has a congenital disease such as thalassemia or sickle cell disease were the success rate of HLA identical sibling cord blood transplant is 90%. The SDCB has now collected 2,531 cord blood units from siblings of patients with Sickle cell disease (28%), thalassemia (5%), other hereditary disorders (15%) and malignant diseases (51%). One hundred and seven patients have been transplanted for thalassemia (24%), Sickle cell disease (21%), malignant disease (33%) and other hereditary disease (22%) representing 4% of the units stored.

Autologous Cord Blood Banks

Public banks collect UCB intended for recipients suffering from hematological diseases (related or unrelated allogeneic transplants), and private banks store newborn blood in order to keep it for that same individual's possible future use (auto transplants). Most commercial banks list many conditions that might be treated in the future by as yet undeveloped stem cell therapies for regenerative medicine. In November 2008, there were more than 100 private banks worldwide accounting for approximately some 1 million units listed on the Parent's guide to cord blood website. www.parentsguidecordblood.org

The growth of public banks is constrained by the public funding on which it depends. In economic terms, the stock of private banks represents the demand of parents who are willing to pay for a service, whereas the public bank stock represents a supply of care services available to its national citizens. Since 75% of the umbilical cords stored worldwide are held by private banks, the imbalance between supply and demand could lead public banks to become dependent on the grafts kept by private banks. While private banks are prohibited in some European countries, such as France and Italy, they are developing quickly in Belgium, Great Britain and Germany. They are growing at a blinding pace in Asia, Australia and the United States. Since 2000, the three main American private banks have increased their graft stock by roughly 40% each year.

Many ethical issues have been raised about the scientific value of commercial CBB for autologous use and their competition with the public CBB [19–21]. In spite of the huge number of private units frozen for autologous use, one case of autologous use has been recently reported in a child with acute leukemia [22]. In its 2004 report on the ethics of cord blood banking, the European Union Group on the Ethics of Science and technology raised serious concerns on the promotion to expectant parents of the future benefits of autologous cord blood banking as a biological insurance to treat a large variety of diseases for which, at present, there is no medical evidence for the validity of the treatment.

The world marrow donor association (WMDA) and other international organizations have expressed their concern about the generalization of autologous cord blood banking and issued the following recommendations: (1) Public donation of cord blood is encouraged, (2) The probability of using one's cord blood is very small and difficult to quantify – as low as 0.04–0.0005% in the first 20 years of life. The lifetime probability of undergoing an autologous HSCT in the US is about 1/400 if the indications do not change much during the next 70 years, (3) There is a concern about the content and viability of cryopreserved cord blood stem cells which could be used for regenerative medicine, (4) There are no data demonstrating the superiority of autologous cord blood cells to adult cells. There was no apparent difference between mesenchymal cells isolated from cord blood or adult bone marrow.

Considering the rapid development of family banks and the intensive research on the use of stem cells for regenerative medicine, Netcord-Eurocord recommends that rules should be established concerning the quality of the unit which should be submitted to the same standards as public unrelated banks which means that the units stored have to be analyzed for serology, number of cells and HLA type, the information to the parents which has to be true and evidence-based without any misunderstanding. Several models of public private banks are under discussion in order to satisfy this increasing demand.

Netcord Organization: Standards and Guidelines

Netcord has developed a detailed set of standards for umbilical cord blood banking which have been submitted to the Foundation for Accreditation of Hemopoietic Cell Therapy (FACT) (USA) for accreditation. These include respective national and international regulatory aspects. Furthermore, a joint system employing most recent Internet technology has been implemented to facilitate rapid allocation of umbilical cord blood units according to histocompatibility and number of nucleated cells within an average time of 48 h. www.netcord.org

Description of Existing Guidelines

The third edition, 2006 of Netcord-FACT International Standards for Cord Blood Collection, Processing, Testing, Banking, Selection and Release is a collaborative effort between Netcord and Foundation for the Accreditation of Cellular Therapy. It is accompanied by a guidance manual directly accessible on the FACT website (2008) [7, 8]. These standards are developed by consensus, based on the best available evidence-based science to the greatest extent possible. They aim to promote quality medical and laboratory practices throughout all phases of cord blood banking to achieve consistent production of high quality umbilical cord blood units for transplantation.

It is divided in sections: A Cord blood bank (CBB) quality management, B CBB operational standards, C cord blood donor management and collections standards, D Cord blood processing standards, E cord blood selection and release standards.

The CBB consists of an integrated team, under a single CBB Director, responsible for the collection, processing, testing banking, selection, release of cord blood units. Following the Netcord-FACT standards; the CBB, each collection facility and each processing facility shall operate in compliance with applicable law, local, and national licensing and registration requirements.

Practical Aspects of Cord Blood Banking

Informed Consent

Cord and placental blood are discarded products that can be used without permission; however, informed consent must be obtained for doing tests on the mother and cord blood. Information on cord blood donation for allogeneic use must be given to the mother long before delivery, and the informed consent process is engaged in advance. This careful collection of medical history allows the opportunity of excluding donors who have a high risk of transmitting infectious or genetic diseases. The families are informed on genetic and infectious tests performed. Donation is free and anonymous without the possibility to retrieve the cells for autologous use.

Collection Techniques

Two techniques are used to collect cord blood: one is collection of the cord blood in the delivery room while the placenta is still in utero and the other is collection of the cord blood in an adjacent room after the delivery. In the first instance, the collection can be done in the delivery room by an obstetrician or a midwife. The advantage of this procedure is that the volume of cells collected is usually higher if the cord is clamped early and the collection is begun immediately; however, this can disrupt the normal process of delivery and is not always feasible. The collection of cord blood after delivery is easier and can be performed by designated personnel;

however, fewer cells may be collected and there may be an increase in the risk of bacterial contamination or clotting.

Infectious Disease Testing

Syphilis and viral tests [including those for human immunodeficiency virus (HIV), hepatitis B (HBV) and C (HCV), and cytomegalovirus (CMV)] are performed on mother's blood. In some countries, the mother is tested for human T-cell lymphotropic virus, type I (HTLV-I), and toxoplasmosis also. Most often, virology tests are not performed immediately on the cord blood; rather, a separate frozen aliquot is kept so that these tests can be performed before a transplant procedure. In some banks cord blood is quarantined until a confirmatory test is performed on the mother 3–6 months after delivery. This should decrease considerably the risk of virus transmission. These risk estimates may be higher in cord blood donors because, unlike regular whole blood donors, cord blood donors are recruited only once. Bacterial infection is also a major issue, but it seems that the incidence of bacterial contamination diminishes with the expertise of the staff in charge of the collection. In all cases, a bacterial culture for anaerobic and aerobic bacteria must be performed and results sent to the transplant physician when the cord blood is delivered for transplantation.

Genetic Disease Testing

The decision to perform tests for genetic diseases should be directed by the family medical background of the family, and the follow-up of the donor. Tests on cord blood are expensive and there is no real consensus on the type and number of tests that should be performed. Also, there are some concerns about notifying the family of the test results some healthy donors might be upset by a diagnosis of genetic disorder without any clinical consequence.

HLA Typing

HLA typing is performed on an aliquot of cord blood. Usually, HLA-A and -B antigens are identified by low resolution allele typing, and HLA-DRB1 is identified by DNA amplification methods. More banks now perform complete molecular high resolution typing for HLA-C, -DQB1, -DPB1, some HLA-A and -B antigens, and other markers. Some banks type the mother for HLA routinely in order to have information on the haplotypes and to control the accuracy of cord blood typing. When donor recipient pairs have been selected most centers perform high resolution typing in both donor and recipient.

Cell Processing

Most banks perform volume reduction before freezing. Many banks freeze cells in a programmable cell freezer. The cord blood is mixed with 10% dimethyl sulfoxide (DMSO) and hydroxyethyl starch (HES) sedimentation for volume reduction and removal of red cells. The thawing technique is well established and aims at removing red cells and DMSO.

Evaluation of the stem progenitor cell content is very important. Several clinical studies have shown that the number of nucleated cells infused correlates with engraftment. There also has been correlation between placental weight, time of clamping, speed of processing, volume collected, and progenitor cell content. Quantification of the cellular content of cord blood is not always easy. Most studies refer to nucleated or mononuclear cells infused per kilogram of body weight before and after thawing of cells. Enumeration of $CD34^+$ cells by flow cytometric analysis is performed routinely by most laboratories, but these results are not always reproducible. Others count the number of granulocyte-macrophage colony-forming units (CFU-GM) in clonogenic assays. There is a lot of variation of these measures among laboratories, which explains why quantification has been a problem. In most studies there is a correlation between the number of nucleated cells, the number of $CD34^+$ cells and CFU-GM and engraftment. The recommended dose is $\geqq 3 \times 10^7$ nucleated cells/kg and $\geq 2 \times 10^5$ CD34 cells/kg infused.

Cryopreservation and Storage

Cryopreservation of unrelated CB units shall be initiated within 48 h of collection, CB unit processing shall be limited to simple dilution and/or volume reduction by depletion of erythrocytes and/or plasma. Any other manipulation shall only be performed using reagents and/or devices approved for that manipulation by the appropriate governmental agency. At a minimum, the following reference samples shall be collected from the unrelated allogeneic, directed allogeneic, or autologous CB unit prior to cryopreservation:

A minimum of two reference aliquots with a minimum volume of 100 µL each sealed and integrally attached to the freezing bag. The contents of each aliquot shall be representative of the CB unit. When a CB unit is initially requested, one segment shall be used for confirmatory typing and should be used for cell viability and/or potency analysis. Cellular aliquots intended for viability or potency analysis shall be stored at –150°C or colder. When cellular aliquots are stored in liquid nitrogen vapour phase at –150°C or colder, the freezers shall be qualified to show that all cellular aliquots are maintained at appropriate temperatures. Cellular aliquots used for purposes other than viability analysis shall be stored at –70°C or colder. Additional cellular aliquots of at least two vials or additional contiguous segments with $1–2 \times 10^6$ nucleated cells per vial or segment.

The following reference samples shall be collected from the infant donor's mother within 7 days before or after the time of CB collection: From the birth mother, serum and/or plasma from non-heparinized samples of at least two vials with a minimum of 1.8 mL each. The serum or plasma should be stored at –70°C or colder. CB units shall be cryopreserved using controlled rate freezing. If an equivalent procedure is used, it shall be validated to maintain equivalent recovery and viability of nucleated cells.

Cryopreservation Standard Operating Procedures shall specify that the following information is recorded for each unit: Total nucleated cell concentration within a defined range, the cryoprotectant, its final concentration, and the duration of cell exposure prior to freezing.

Each CB unit shall be maintained in quarantine storage until the CBB Director or designee has approved the release of the CB unit from quarantine status based upon review of maternal communicable disease risk history, other medical history, maternal test results, and CB unit sterility test results as required under Applicable Law. Records shall indicate when a CB unit was released from quarantine into permanent storage.

Unrelated allogeneic CB units shall be quarantined if the unit or maternal samples have positive or indeterminate screening test results for communicable disease.

The following tests shall be performed on a sample from each CB unit: From the final CB unit at the end of processing prior to cryopreservation: Total nucleated cell count, Nucleated red blood cell count, Total number of CD34 cells. Viability and/or potency from the final CB unit are measured by TNC, CD34 cells, and/or CFU. Microbial cultures of the CB unit or product obtained after processing prior to cryopreservation using a system permissive for the growth of aerobic and anaerobic bacteria and fungi, ABO group and Rh type, Human leukocyte antigen (HLA) type, HLA-A, B, and DRB-1 loci shall be determined. HLA-C and DQB should be determined. HLA Class I and Class II typing shall be performed by DNA-based methods. For unrelated allogeneic CB units, this typing shall be performed before listing the CB unit for search. As a minimum, DNA high resolution molecular typing shall be used for Class II DRB-1 typing before release of the CB unit to the Clinical Program.

Before release the following tests will be performed: Hemoglobinopathy screening, CFU total cell number from the final CB.

Before cryopreservation, a blood count with differential should be performed on all CB units and parameters for neutrophils, lymphocytes, and platelets shall be defined. Prior to release for transplantation, each CB unit should be tested for evidence of infection by at least the following communicable disease agents using licensed tests are available according to Applicable Law: Human immunodeficiency virus, type 1, Human immunodeficiency virus, type 2, Hepatitis B virus, Hepatitis C virus, Human T cell lymphotropic virus, type I, Human T cell lymphotropic virus, type II. *Treponema pallidum* (syphilis) and any additional agents required by Applicable Law at the time of release of the CB unit.

Ethical and Legal Aspects

Cord blood can be collected for infusion into an unrelated or a related recipient and, more hypothetically, for autologous use. For unrelated transplantation, the mother must be aware that the donation is anonymous and free, and that there will be no guarantee that it will be possible to retrieve the cord blood if it is needed later for family or autologous use.

Clinical Experience with Related and Unrelated Umbilical Cord Blood Transplantation

The CBB are obliged to maintain sufficient critical outcome data to ensure that the procedures in use in the CBB consistently provide a safe and effective product. While it is understood that a CBB is not in control of a Clinical Program's responsiveness in providing outcome data, the CBB should make it clear in an agreement with the Clinical Program that it is required to obtain this information for analysis of quality, safety, and efficacy and demonstrate diligence in obtaining a high percentage of at least Day 100 and 1-year outcome data. In order to obtain validated data most banks send their data to registries: Eurocord for Europe and NMDP for the USA. Common questionnaires are used each bank, they notify the registry at time an unit is released, and the registry contacts the transplant center and collect the clinical data which are sent back to the bank. In addition some registries e.g. Eurocord perform analysis on indications, criteria of donor choice, comparison with other sources of hematopoietic stem cells. These data have been critical in the development of the direction in which cord blood transplantation have developed.

A survey of the International Bone Marrow Transplant registry (CIBMTR) estimates that after 1998, 20% of stem cell transplants performed in young patients (<20 years old) are cord blood transplants (IBMTR Newsletter). In Japan, nowadays approximately 50% of HSCT from unrelated donors are being performed with cord blood cells.

Eurocord www.eurocord.org is an international registry which operates on behalf of the European Blood and Marrow Transplant group (EBMT), which includes European and non European centers (more than 180 transplants centers in 35 countries), all performing either related or unrelated cord blood transplants. It works in close collaboration with EBMT and Netcord banks (www.netcord.org) to collect clinical data and follow patients transplanted in or outside Europe with Netcord units. Once a year, the Eurocord and EBMT data bases are carefully checked to detect overlaps and discrepancies between data reported, and to verify the compliance of the transplant centers. This ensures that all consecutive transplants at the EBMT centers are registered in the Eurocord data base. Regarding non-EBMT centers, they are asked to report their cord blood transplants if the cord blood units came from Netcord banks. Data concerning patient and disease characteristics and transplant outcomes are collected by standardized questionnaires. Submitted data are reviewed by two physicians and computerized error checks are performed to ensure data quality. Eurocord participates to the Fact Netcord accreditation by giving regularly outcome data to the banks in order to fulfil requirements on the quality of the transplant. An agreement has been signed with NMDP (National Marrow Donor Program) and CIBMTR (Center for International Bone Marrow Transplant Research) to share data outcome in Europe and USA. Once a year, Eurocord data base and EBMT data base are carefully checked to detect overlaps, and discrepancies between data reported and to verify the compliance of the transplant centers. This ensures that all consecutive transplants for the EBMT centers are registered in Eurocord data base. Regarding the non-EBMT centers, they are asked to report their cord blood transplants if cord blood units came from Netcord banks. Data concerning patient and disease characteristics and transplant outcomes are collected by standardized questionnaires. Submitted data are reviewed by two physicians and computerized error checks are performed to ensure data quality.

Thanks to this collaboration, from 1988 to October 2008, 4,875 cord blood transplantations have been reported to the Eurocord registry from 233 European transplant centers and 197 transplant centers from other countries. Five hundred and two transplants have been reported using related donors (majority of HLA identical sibling donors) mainly for children with malignant and non malignant disorders and 4,373 have been performed in the unrelated transplant setting for children (2,901) and adults (1,882). During the last 3 years, the number of unrelated UCBT reported to Eurocord has increased to more than 300 transplants/year and since 2004 the number of adults transplanted with cord blood has overcome the number of UCBT performed in children (Eurocord registry unpublished data).

In order to promote information Eurocord has recently launched a new European program entitled: an On-line CME programme in cord blood technology and transplantation for providing a learning tool on the scientific, technical, clinical, regulatory aspects of cord blood, easily accessible at a time and language convenient for users www.eurocord-ed.org.

Clinical Results of Cord Blood Transplantation from Eurocord Registry

Cord blood transplantation has been used to treat a variety of genetic, hematologic, immunologic, metabolic and oncologic disorders.

Influence of Cell Dose and HLA on Transplantation Outcome

One of the major advantages of using cord blood instead of bone marrow is the diminution of alloreactivity of cord blood lymphocytes resulting in the diminution of acute and chronic GVH after related or unrelated transplant. Optimizing cord blood donor selection might further improve results of unrelated cord blood transplant. In order to study the association of nucleated cell dose (NC) and number of HLA disparities with outcomes after UCBT, Eurocord registry collected and analyzed 550 UCBT recipients for hematological malignancies. Multivariate analysis of factors associated with outcome in malignant diseases is shown in Table 3.4.

The current recommendations are currently to choose:

- Cord blood units with ≤ 2 HLA disparities and $>3 \times 10^7$ nucleated cells/kg or $\geq 2 \times 10^5$ CD34+ cells/kg
- In non malignant disease where the risk of rejection is higher the dose should be increased and one must avoid units with less than 3.5×10^7 NC/kg and 2 or more HLA incompatibility. If there is no single unit with these characteristics look for two units with a combined total dose of $\geq 3 \times 10^7$ NC/kg and if possible not more than 1 HLA difference between the 2 units and the patient.

Table 3.4 Factors associated with favourable outcomes in patients with malignant diseases (multivariate analysis)

Neutrophil engraftment • HLA 6/6 or 5/6 • Early and intermediate disease • Cells infused >2 × 10^7 per kg	Platelet engraftment • HLA 6/6 or 5/6 • Early and intermediate disease • Cells infused >2 × 10^7 per kg • Patients negative CMV serology
Transplant related mortality • HLA 6/6 or 5/6 • Early and intermediate disease • Cells infused >2 × 10^7 per kg	Relapse • HLA 4/6 or 3/6 • Early or intermediate disease

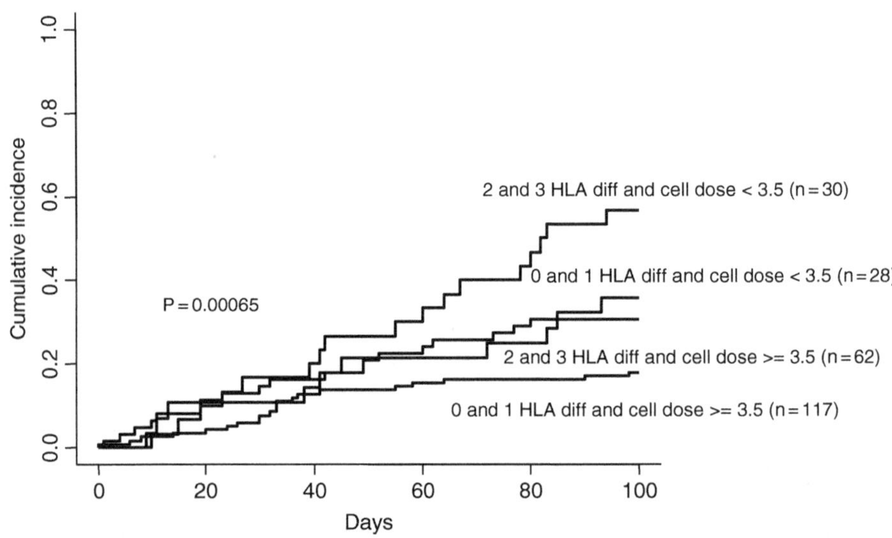

Fig. 3.3 Role of HLA and cell dose on transplant related mortality after unrelated cord blood transplant for malignant disease

These data strongly suggest that UCB is an acceptable alternative to matched unrelated BM in children, and support the start of a simultaneous search for BM and UCB unrelated donors. The final selection of unrelated donor BM versus UCB should be based on the urgency of the transplant, and the characteristics of the BM and UCB unrelated donor such as cell dose and HLA compatibility. For those children requiring an urgent transplant, generally less than 3 months, UCB seems advantageous. Moreover, with the aim to find more closely matched CB grafts, cord blood banks should increase their inventories.

The results of 4 comparative studies and the meta-analysis, gathered together showed that (1) UCBT is feasible in adults when a cord blood unit contains a higher number of cells and should be considered an option as an allogeneic stem cell source for patients lacking a HLA matched bone marrow donor. (2) Despite increased HLA disparity, UCB from unrelated donors offers sufficiently promising results to matched UBM in adults with hematological malignancies leading to the conclusion, as in children, that the donor search process for BM and UCB from unrelated donors should be started simultaneously especially in patients with acute leukemia where the time factor is crucial.

Conclusion

Umbilical cord blood has emerged as an appealing alternative source of hematopoietic stem cells for transplantation. Although many issues remain uncertain and greater experience will be required to determine clearly the relative merits of UCBT compared with BMT, all available data suggest that unrelated donor UCBT should be considered an acceptable option in children and adults with hematologic malignancies or non malignant disorder for whom an HLA-matched BM unrelated donor is not readily available. The choice of the stem cell graft will depend on the urgency of the transplant, the cell dose and number of HLA disparities. Hopefully, the increase number of high quality cord blood units, current research approaches and the greater experience of transplant centers on UCBT will improve outcomes and will provide successful therapy to a larger number of patients in need.

References

1. Gluckman E, Broxmeyer HE, Auerbach AD et al (1989) Hematopoietic reconstitution in a patient with Fanconi's anemia by means of umbilical-cord blood from an HLA-identical sibling. N Engl J Med 321:1174–1178
2. Barker JN, Krepski TP, DeFor T et al (2002) Searching for unrelated donor hematopoietic stem cell grafts: availability and speed of umbilical cord blood versus bone marrow. Biol Blood Marrow Transplant 8:257–260
3. Frassoni F, Podesta M, Maccario R, Giorgiani G, Zecca M, Bacigalupo A, Piaggio G, Locatelli F (2003) Cord blood transplantation provides better reconstitution of hematopoietic reservoir compared with bone marrow transplantation. Blood 102:1138–1141
4. Rocha V, Wagner JE, Sobocinski KA et al (2000) Graft-versus-host disease in children who have received a cord blood or bone marrow transplant from an HLA-identical sibling. N Engl J Med 342:1846–1854
5. Cairo MS, Wagner EL, Fraser J, Cohen G, van de Ven C, Carter SL, Kernan NA, Kurtzberg J (2005) Characterization of banked umbilical cord blood hematopoietic progenitor cells and lymphocyte subsets and correlation with ethnicity, birth weight, sex, and type of delivery: a cord blood transplantation (COLBT) study report. Transfusion 45:856–866
6. Ballen KK, Kurtzberg J, Lane TA, Lindgren BR, Miller JP, Nagan D, Newman B, Rupp N, Haley NR (2004) Racial diversity with high nucleated cell counts and CD34 counts achieved in a national network of cord blood banks. Biol Blood Marrow transplant 10:269–275
7. Wernet P (2008) The Netcord inventory and use [Netcord web site]. Oct 2008. Available at: https://office.de.netcord.org/inventory.gif
8. NETCORD 3rd (2006) FACT - NETCORD - International Standards for cord blood collection, processing, testing, banking, selection, and release, 3rd edition, Dec 2006
9. NETCORD CBB 2008: NETCORD CBB Guidance Manual 03/03/08 Guidance derived from meetings with the Scientific Experts Committee and the European Commission

10. Rubinstein P, Dobrila L, Rosenfield RE, Adamson JW, Migliaccio G, Migliaccio AR, Taylor PE, Stevens CE (1995) Processing and cryopreservation of placental/umbilical cord blood for unrelated bone marrow reconstitution. PNAS 92:10119–10122
11. Jones J, Stevens CE, Rubinstein P, Robertazzi RR, Kerr A, Cabbad MF (2003) Obstetrics predictors of placental/umbilical cord blood volume for transplantation. Am J Obstet Gynecol 188:503–509
12. Rebulla P, Lecchi L, Porretti L, Poli F, Ratti I, Mozzi F, Sirchia G (1999) Practical placental blood banking. Transfusion Medicine Reviews 13: 205–226
13. Solves P, Mirabet V, Planelles D, Carbonnell-Uberos F, Roig R (2008) Influence of volume reduction and cryopreservation methodologies on quality of thawed umbilical cord blood units for transplantation. Cryobiology 56:152–158
14. Kodera Y (2008) The Japan marrow donor program, the Japan cord blood bank network and the Asia blood and marrow registry. Bone Marrow Transplant 42(1):56
15. Davey S, Armitage S, Rocha V, Garnier F, Brown J, Brown CJ, Warwick R, Fehily D, Watt S, Gluckman E, Vora A, Contreras M, Navarrete C (2004) The London Cord blood Bank: analysis of banking and transplantation outcome. Br J Haematol 125:358–365
16. Howard DH, Meltzer D, Kollman C, Maiers M, Logan B, Gragert L, Setterholm M, Horowitz MM (2008) Use of cost effectiveness analysis to determine inventory size for a national cord blood bank. Med Dec making 28:243–253
17. Reed W, Smith R, Dekovic F, et al (2003) Comprehensive banking of sibling donor cord blood for children with malignant and non malignant disease. Blood 101:351–357
18. Lubin B, Shearer WT American academy of pediatrics Section on Hematology/Oncology (2007) Cord blood banking for potential future transplantation. Pediatrics 119:165–170
19. ASBMT position statement (2008) Collection and preservation of cord blood for personal use. Biol Blood Marrow Transplant 14:364–362
20. Ballen KK, Barker JN, Stewart SK, Greene MF, Lane TA (2008) Collection and preservation of cord blood for personal use. Biol Blood Marrow Transplant 14:356–363
21. Sullivan MJ (2008) Banking on cord blood stem cells. Nat Rev Cancer 8:554–563
22. Hayani A, Lampeter E, Viswanatha D, Morgan D, Salvi SN (2007) First report of autologous cord blood transplantation in the treatment of a child with leukemia. Pediatrics 119:296–300
23. Gluckman E, Rocha V, Boyer-Chammard A et al (1997) Outcome of cord blood transplantation from related and unrelated donors. Eurocord Transplant Group and the European Blood and Marrow Transplantation Group. N Engl J Med 337:373–381
24. Rubinstein P, Carrier C, Scaradavou A et al (1998) Outcomes among 562 recipients of placental-blood transplants from unrelated donors. N Engl J Med 339:1565–1577
25. Wagner JE, Barker JN, DeFor TE et al (2002) Transplantation of unrelated donor umbilical cord blood in 102 patients with malignant and nonmalignant diseases: influence of CD34 cell dose and HLA disparity on treatment-related mortality and survival. Blood 100:1611–1618
26. WMDA (2007) World marrow donor association guidelines for use of HLA nomenclature and its validation in the data exchange among hematopoietic stem cell donor registries and cord blood banks. Bone Marrow Transplant 39:737–741
27. Kamani N, Spellman S, Hurley CK, Barker JN, Smith FO, Oudshoorn M, Bray R, Smith A, Williams TM, Logan B, Eapen M, Anasetti C, Setterholm M, Confer DL, National Marrow Donor Program (2008) State of the art: HLA matching and outcome of unrelated donor unrelated donor umbilical cord blood transplants. Biol Blood Marrow Transplant 14:1–6

Chapter 4
Banking of Corneas

W. John Armitage

Cornea is one of the oldest transplants of an anatomically and physiologically intact, viable tissue with the first successful full-thickness corneal allograft being attributed to Eduard Zirm in 1905 [1]. The patient was a farm labourer who had suffered bilateral lime burns, a condition that even today has a poor prognosis, which makes Zirm's success all the more remarkable. At that time, it was not only thought that tissue taken from a deceased donor would be harmful to the recipient but there was no satisfactory means of tissue preservation. The corneal graft was therefore taken from a child's eye that had just been enucleated following an injury and used immediately. By contrast, corneas are now stored routinely in eye banks for up to 4 weeks and distributed nationally and internationally for elective transplant surgery. Approximately 35,000 corneal transplants are carried out in the USA annually with perhaps a further 20,000 each year in Europe. While cornea is by far the most frequently transplanted ocular tissue, sclera and limbal tissue are also used therapeutically for, respectively, reconstructive surgery and the treatment of ocular surface disease. Cryopreserved amniotic membrane is also used to treat ocular surface disease [2–4].

Eye Donation

Eyes are typically retrieved up to 24 h after death. In the UK, the whole eye is enucleated with sterile, single-use instruments; but in some countries an in situ excision of the cornea with a 2–4 mm rim of sclera (corneoscleral disc) is performed, leaving the rest of the eye intact in the donor. Restoration of the donor's appearance after eye or corneal retrieval is an important final step of the eye donation procedure.

Some donor selection criteria for corneal transplantation differ from other types of tissue transplant owing mainly to the avascularity of the cornea. It is accepted, therefore, that cancers, with the exception of haematological or ocular malignancies,

W.J. Armitage (✉)
CTS Bristol Eye Bank, University of Bristol, Bristol Eye Hospital, Bristol BS1 2LX, UK
e-mail: w.j.armitage@bristol.ac.uk

G. Galea (ed.), *Essentials of Tissue Banking*, DOI 10.1007/978-90-481-9142-0_4,

do not necessarily preclude eye donation. Depending on the method of corneal storage, infections and bacteraemia may also be accepted. Apart from these exceptions for cornea, which do not apply to limbal tissue or sclera since these are vascularized, the main exclusion criteria are similar to other tissues. Provided the corneal endothelium is examined to ensure an adequate number of cells, there is no need to set an upper age limit for cornea donation.

Corneal Transplantation

The cornea forms part of the outer coat of the eye, merging with the sclera at the corneal limbus (Fig. 4.1). It is tough, having both to withstand the intraocular pressure and protect the delicate internal structures of the eye. The human cornea is approximately 0.55 mm thick and is bounded on the outer anterior surface by a stratified epithelium 5–7 cells thick and on the inner posterior surface by a monolayer

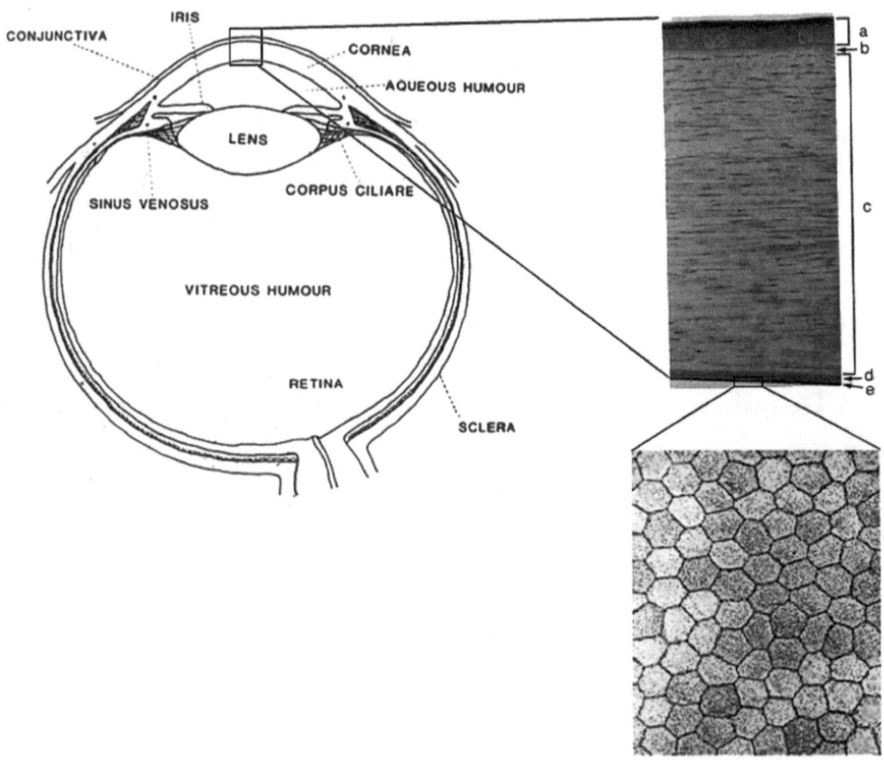

Fig. 4.1 Morphology of eye, cornea and corneal endothelium. Key to corneal section: a, epithelium; b, Bowman layer; c, stroma; d, Descemet membrane; e, endothelium. The posterior surface of the endothelium shows a mosaic of closely apposed, mainly hexagonal cells. (Reproduced from Hogan et al. [41] with permission.)

of closely apposed endothelial cells (Fig. 4.1). The bulk of the cornea comprises a collagenous stroma containing fibroblastic cells, the keratocytes.

The cornea is the major refracting element of the eye and this function critically depends on its transparency [5]. Accordingly, normal cornea allows transmission of between 86 and 94% of light in the visible spectrum. This remarkable degree of clarity is a result of the long-range ordering achieved by the highly structured organization of the uniformly-sized stromal collagen fibrils embedded in a proteoglycan matrix [6]. These fibrils run parallel to each other in sheets (lamellae) that stretch across the width of the cornea with successive lamellae being arranged orthogonally. The fixed negative charge on the proteoglycans draws water and solutes into the stroma from the aqueous humour. In the absence of blood vessels, this influx of solutes is essential for the nutrition of the keratocytes. The consequent influx of water is countered by the endothelium, which forms a passive, albeit leaky, barrier and actively pumps ions, including bicarbonate, from the stroma to the aqueous, which induces an efflux of water thereby controlling stromal hydration [7, 8]. If either the endothelial barrier properties or the ion pumping are disrupted, the stroma will become oedematous with a consequent loss of transparency. Human endothelium has only a limited proliferative capacity and, as a result, there is a continuous decline in endothelial cell density throughout life [9].

Approximately 40% of corneal transplants in the UK are to treat loss of transparency as a consequence of endothelial disease, principally Fuchs' endothelial dystrophy and bullous keratopathy – the latter being a consequence of previous ocular surgery. The other major indication for transplantation is severe visual impairment caused by a change in corneal shape. In this tissue matrix disease, keratoconus, the cornea progressively thins and its surface becomes conical instead of spherical. Other indications for transplantation include infections, such as herpes simplex keratitis, scarring, opacification and, increasingly, regraft.

In full-thickness corneal grafts (penetrating keratoplasty, PK) a disc of diseased cornea, typically 7.5 mm in diameter, which includes epithelium, stroma and endothelium, is removed and replaced by a similarly sized graft of healthy corneal tissue from a donor (Fig. 4.2). Partial thickness grafts (lamellar keratoplasty) are also used where just the diseased part of the cornea, rather than all three layers, is replaced [10]. In endothelial keratoplasty (EK), diseased endothelium is replaced by donor endothelium supported on either a thin layer of stroma approximately 150 μm thick or even just on Descemet membrane. The advantages of EK over PK are faster visual rehabilitation for the patient and less astigmatism, which is a common postoperative complication of PK; however, it is uncertain at the time of writing whether long-term EK survival is similar to PK. Both PK and EK critically depend on an allograft with an intact, functioning endothelial cell layer. Assessment of the endothelium is therefore important to ensure an adequate number of cells and good morphology of this cell monolayer for these types of allograft.

Deep anterior lamellar keratoplasty (DALK) is an alternative to PK for treating keratoconus with the purported advantage that the recipient's own healthy endothelium is not replaced but remains intact. Other anterior lamellar grafts can be used to replace scarred or opacified stromal tissue. Clearly, there is no requirement for the

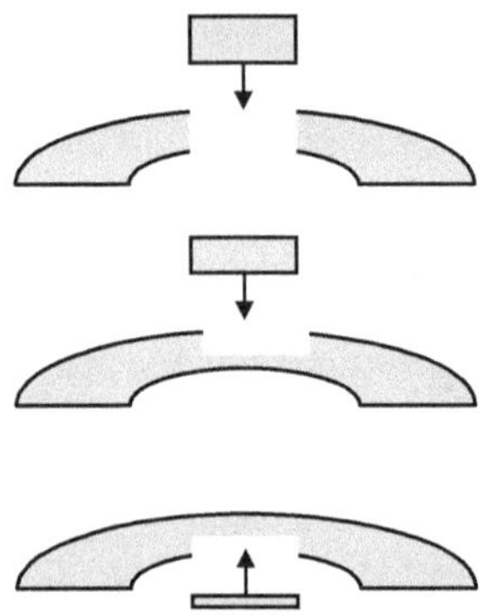

Fig. 4.2 Types of corneal transplant. Penetrating keratoplasty is a full-thickness graft where a central disc of host cornea (typically 7.5 mm) is removed and replaced with tissue from a donor cornea. Where the host endothelium is healthy but the stroma is misshapen and thinned, a partial thickness graft (DALK) leaves the host endothelium intact. For endothelial disease, replacement with healthy endothelium on a thin layer of stroma or just on Descemet membrane is an option (EK)

corneas used for these types of allografts to have an intact endothelium. However, corneas for DALK are often supplied with a good endothelium in case the surgeon needs to switch to PK during the operation owing to accidental perforation and breach of the anterior chamber during the lamellar dissection of the patient's cornea.

Corneal Storage

Hypothermia and organ culture are the two principal methods used for storing corneas. Cryopreservation has been used to a very limited extent in the past but is currently rarely used and then only for emergency grafts to save an eye rather than to improve vision [11].

Hypothermia

It was not until the 1930s that Filatov pioneered the use of corneas from deceased donors and the storage of eyes in pots (moist chambers) in ice for several days [12]. This method continued in use until the early 1970s when removal of the corneoscleral disc from the eye and storage in tissue culture medium containing 5% dextran (McCarey-Kaufman medium, M-K medium) was introduced [13]. The corneas were refrigerated at 4°C for 2–4 days compared with the 24–48 h storage time for whole eyes in moist chambers.

The principle underpinning hypothermic storage is the reduction in rates of chemical reactions with falling temperature. For biological reactions there is a 2–3 fold reduction in rates for every 10°C fall in temperature, which means that the energy demands of cells are markedly lower at 4°C than at normothermia. There are, however, several limitations to the maximum storage period [14]. In particular, metabolism is reduced but not completely suppressed, which means that there is still a demand for high energy compounds such as ATP. The ability of cells to generate ATP at these low temperatures is compromised, which leads to an overall loss of high energy compounds. Ion pumps are also suppressed, leading to altered ionic balance and cellular oedema and cellular acidosis. There are also possible phase changes in membrane lipids and generation of reactive oxygen species (ROS) by the Fenton reaction.

The hypothermic storage medium currently most widely used is Optisol [15, 16]. This solution contains both dextran and chondroitin sulphate to help control stromal hydration and provides up to 14 days of storage; however, most eye banks do not store beyond 7–10 days owing to poor epithelial preservation. There are other media available that aim to support the reduced cellular metabolism, counter damage from ROS and support cell membrane repair [17–19].

Organ Culture

Organ culture of corneas was introduced in the USA in the early 1970s and was based on culture techniques that had been developed for skin [20, 21]. Owing to its perceived complexity, the occurrence of stromal oedema during storage and potential problems of bacterial and fungal contamination, North American eye banks favoured hypothermic storage in M-K medium. However, the technique was further developed in Denmark and became the method of choice for many eye banks in Europe [22–25].

A key element to the method is the cleaning of the ocular surface before excision of the corneoscleral disc by rinsing in sterile saline and immersion of the eye in Povidone-iodine (PVP-I) solution [26]. The corneoscleral disc is then suspended in organ culture medium for up to 4 weeks. In the UK, corneas are kept in Eagle's minimum essential medium with Earle's salts containing 26 mM sodium bicarbonate, HEPES buffer, 2% fetal bovine serum and antibiotics (penicillin and streptomycin) and an antimycotic (amphotericin B) [27]. This medium is not changed during the organ culture period and is sufficient to maintain the integrity of the epithelium and endothelium [28]. Alternatively, some eye banks do change the medium every 1 or 2 weeks and there have even been reports of successful corneal transplants with tissue stored for up to 7 weeks with changes of medium [29]. A wide range of different media have been used for corneal organ culture with seemingly similar clinical outcomes. Recently, there have been attempts to remove serum and constituents of animal origin with encouraging results [30].

In the organ culture method as applied in the UK, a sample of the medium is taken after 7 days of culture to test for bacterial and/or fungal contamination. Three days before the scheduled date of transplant, the corneal endothelium is examined by transmitted light microscopy after staining with trypan blue to identify dead or missing cells and hypotonic sucrose that renders the cell borders visible.

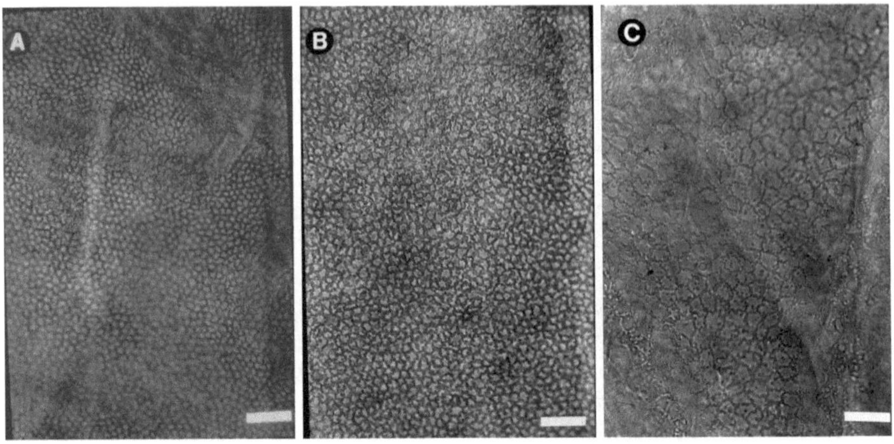

Fig. 4.3 Corneal endothelium after organ culture and staining with trypan blue and hypotonic sucrose (bar = 100 μm). (**a**) Cornea from an 18 year old donor after 23 days of storage, cell density >3,000 cells/mm^2. (**b**) Cornea from a 75 year old donor after 29 days of storage, cell density >2,500 cells/mm^2. (**c**) Cornea from an 84 year old donor after 26 days of storage, cell density <1,000 cells/mm^2 (unsuitable for penetrating keratoplasty or endothelial keratoplasty). (Reproduced from Armitage [42] with permission)

In the UK, a minimum endothelial cell density of 2,200 cells/mm^2 is considered acceptable for corneas to be used for PK or EK (Fig. 4.3). Corneas are then transferred to organ culture medium containing 5% dextran in order to reverse the stromal oedema that occurs during organ culture storage. A further sample of medium is taken after 24 h for microbiological testing and then the cornea is despatched to the recipient hospital. Up to 4 days storage is permitted in the dextran medium.

The donor's cause of death is the main factor influencing whether corneas are lost during organ culture through microbial contamination [27]. Virtually every eye coming to the eye bank will be carrying bacteria and/or fungi on the ocular surface consequent to the loss of tear film and blinking, but it would appear that this microbial load varies with the cause of death. It is not surprising, therefore, that corneas from donors that died with infections are at greater risk of becoming contaminated during organ culture. So far as the likelihood of the corneal endothelium meeting the minimum criteria for PK or EK is concerned, the major influencing factor is increasing donor age. Up to the age of 60 years, approximately 10% of corneas fail to meet the minimum endothelial cell density of 2,200 cells/mm^2, but this rises to more than 30% for donors over the age of 80 years [27].

Five year graft survival for PK in the UK with organ cultured corneas is 70% (95% CI 68–72) [31]. The indication for transplantation has the largest influence on survival, which ranges from 91% for keratoconus to less than 60% for bullous keratopathy. No donor or storage factors have been found to influence graft survival, which suggests that the donor selection criteria, storage time and endothelial assessment are all appropriate. These clinical follow-up data provide direct validation of the efficacy of the organ culture method (unpublished results based on analysis of NHS Blood and Transplant data).

Limbal Stem Cells

The corneal epithelium forms a protective barrier on the anterior surface of the cornea. In conjunction with the tear film, the epithelium provides a smooth anterior corneal surface, which is essential for the transmission of light and thus normal vision. Unlike the endothelium, the epithelium is constantly being renewed. The progeny of a population of slow-cycling stem cells residing in the basal limbal epithelium become transient amplifying cells (TAC) that migrate to the basal layer of the corneal epithelium where they divide and thus maintain the epithelial cell mass [32, 33]. As the TAC divide, their progeny move anteriorly becoming post-mitotic and, eventually, they form a superficial layer of terminally differentiated cells. Failure of the limbal stem cell population compromises the integrity of the epithelium, which results in ocular surface disease that can be painful and difficult to treat, causing severe visual impairment. Attempts to restore vision by corneal transplantation are almost certain to fail. This limbal stem cell deficiency (LSCD) can be treated by replacement of the stem cells by transplantation of limbal tissue or the application of epithelial cell sheets expanded in vitro from limbal tissue explants onto the ocular surface [34].

If the disease is unilateral, the best chance of successful treatment is with an autograft taken from the healthy eye. A small limbal biopsy can then be used as a source of stem cells that are expanded in vitro on a support such as fibrin or amniotic membrane [34]. If the disease is bilateral, then one option is to use allograft limbal tissue or cells, but this requires immunosuppression in an attempt to prevent allograft rejection and, even so, there is little evidence that such allografted cells persist in the long term [35]. Alternatively, other sources of autologous cells, such as oral mucosa, are being explored [36].

Despite the therapeutic use of limbal tissue grafts, there is still no single marker for corneal epithelial stem cells that can be used specifically to identify and enrich this cell population, which forms just a few percent of the cells in the limbal basal epithelium [37]. However, combinations of several immunohistochemical markers do appear to localize to limbal stem cells, such as positive staining for transcription factor p63 and the ATP binding cassette transporter ABCG2, but negative staining for cytokeratins K3/K12 and for connexin 43 [33]. Nonetheless, limbal epithelial cells can be successfully expanded in vitro and there is currently a focus on improving the culture technique, for example to avoid the need for 3T3 cell feeder layers, bovine serum and other medium constituents of animal origin [38, 39]. Alternatives to amniotic membrane and fibrin as supports for epithelial cell sheets are also being investigated [40].

In summary, full-thickness corneal transplants are still the most commonly performed ocular tissue transplant but partial-thickness grafts are becoming far more frequent, especially endothelial keratoplasty. The treatment of ocular surface disease with limbal tissue or epithelial cell grafts is far less common and still the subject of continuing research into improving the cultivation of epithelial cell sheets and exploring other sources of cells that could be used for autologous grafts in patients with bilateral ocular surface disease.

References

1. Armitage WJ, Tullo AB, Larkin DFP (2006) The first successful full-thickness corneal transplant: a commentary on Eduard Zirm's landmark paper of 1906. Br J Ophthalmol 90:1222–1223
2. Tseng SC (2001) Amniotic membrane transplantation for ocular surface reconstruction. Biosci Reports 21:481–489
3. Maharajan VS, Shanmuganathan V, Currie A, Hopkinson A, Powell-Richards A, Dua HS (2007) Amniotic membrane transplantation for ocular surface reconstruction: indications and outcomes. Clin Exp Ophthalmol 35:140–147
4. Kruse FE, Cursiefen C (2008) Surgery of the cornea: corneal, limbal stem cell and amniotic membrane transplantation. Dev Ophthalmol 41:159–170
5. Klyce SD, Beuerman RW (1998) Structure and function of the cornea. In: Kaufman HE, Barron BA, McDonald MB (eds) The cornea. Butterworth-Heinemann, Boston, pp 3–50
6. Maurice DM (1984) The cornea and sclera. In: Davson H (ed) The eye, vol 1b. Vegetative physiology and biochemistry. Academic Press, Orlando, FL, pp 1–158
7. Dikstein S, Maurice DM (1972) The metabolic basis to the fluid pump in the cornea. J Physiol 221:29–41
8. Maurice DM (1972) The location of the fluid pump in the cornea. J Physiol 221:43–54
9. Armitage WJ, Dick AD, Bourne WM (2003) Predicting endothelial cell loss and long-term corneal graft survival. Invest Ophthalmol Visual Sci 44:3326–3331
10. Alio JL, Shah S, Barraquer C, Bilgihan K, Anwar M, Melles GRJ (2002) New techniques in lamellar keratoplasty. Curr Opin Ophthalmol 13:224–229
11. Armitage WJ (2008) Developments in corneal preservation. In: Reinhard T, Larkin F (eds) Cornea and external eye disease. Springer, Berlin, pp 101–109
12. Filatov VP (1935) Transplantation of the cornea. Arch Ophthalmol 13:321–347
13. McCarey BE, Kaufman HE (1974) Improved corneal storage. Invest Ophthalmol Visual Sci 13:165–173
14. Fuller BJ (1991) The effects of cooling on mammalian cells. In: Fuller BJ, Grout BWW (eds) Clinical applications of cryobiology. CRC Press, Boca Raton, pp 1–22
15. Lindstrom RL, Kaufman HE, Skelnik DL, Laing RA, Lass JH, Musch DC et al (1992) Optisol corneal storage medium. Am J Ophthalmol 114:345–356
16. Smith TM, Popplewell J, Nakamura T, Trousdale MD (1995) Efficacy and safety of gentamicin and streptomycin in Optisol-GS, a preservation medium for donor corneas. Cornea 14:49–55
17. Chen CH, Rama P, Chen SC, Sansoy FN (1997) Efficacy of organ preservation media enriched with nonlactate-generating substrate for maintaining tissue viability: a transplantation study. Transplantation 63:656–663
18. Serbecic N, Beutelspacher SC (2005) Anti-oxidative vitamins prevent lipid-peroxidation and apoptosis in corneal endothelial cells. Cell Tissue Res 320:465–475
19. Steinhardt RA, Alderton JM (2006) Poloxamer 188 enhances endothelial cell survival in bovine corneas in cold storage. Cornea 25:839–844
20. Summerlin WT, Miller GE, Harris JE, Good RA (1973) The organ-cultured cornea: an in vitro study. Invest Ophthalmol Visual Sci 12:176–180
21. Doughman DJ, Harris JE, Mindrup E, Lindstrom RL (1982) Prolonged donor cornea preservation in organ culture:long-term clinical evaluation. Cornea 1:7–20

22. Sperling S (1979) Human corneal endothelium in organ culture. The influence of temperature and medium of incubation. Acta Ophthalmol Scand 57:269–276
23. Sperling S (1978) Early morphological changes in organ cultured human corneal endothelium. Acta Ophthalmol Scand 56:785–792
24. Pels E, Schuchard Y (1983) Organ-culture preservation of human corneas. Documenta Ophthalmol 56:147–153
25. Maas-Reijs J, Pels E, Tullo AB (1997) Eye banking in Europe 1991–1995. Acta Ophthalmol Scand 75:541–543
26. Pels E, Vrensen GF (1999) Microbial decontamination of human donor eyes with povidone-iodine: penetration, toxicity, and effectiveness. Br J Ophthalmol 83:1019–1026
27. Armitage WJ, Easty DL (1997) Factors influencing the suitability of organ-cultured corneas for transplantation. Invest Ophthalmol Visual Sci 38:16–24
28. Crewe JM, Armitage WJ (2001) Integrity of epithelium and endothelium in organ-cultured human corneas. Invest Ophthalmol Visual Sci 42:1757–1761
29. Ehlers H, Ehlers N, Hjortdal JO (1999) Corneal transplantation with donor tissue kept in organ culture for 7 weeks. Acta Ophthalmol Scand 77:277–278
30. Thuret G, Manissolle C, Campos-Guyotat L, Guyotat D, Gain P (2005) Animal compound-free medium and poloxamer for human corneal organ culture and deswelling. Invest Ophthalmol Visual Sci 46:816–822
31. Transplant activity in the UK 2006–2007. NHS Blood and Transplant (2007)
32. Pellegrini G, Golisano O, Paterna P, Lambiase A, Bonini S, Rama P et al (1999) Location and clonal analysis of stem cells and their differentiated progeny in the human ocular surface. J Cell Biol 145:769–782
33. Schlötzer-Schrehardt U, Kruse FE (2005) Identification and characterization of limbal stem cells. Exp Eye Res 81:247–264
34. Rama P, Bonini S, Lambiase A, Golisano O, Paterna P, De Luca M et al (2001) Autologous fibrin-cultured limbal stem cells permanently restore the corneal surface of patients with total limbal stem cell deficiency. Transplantation 72:1478–1485
35. Williams KA, Brereton HM, Aggarwal R, Sykes PJ, Turner DR, Russ GR et al (1995) Use of DNA polymorphisms and the polymerase chain reaction to examine the survival of a human limbal stem cell allograft. Am J Ophthalmol 120:342–350
36. Ang LPK, Nakamura T, Inatomi T, Sotozono C, Koizumi N, Yokoi N et al (2006) Autologous serum-derived cultivated oral epithelial transplants for severe ocular surface disease. Arch Ophthalmol 124:1543–1551
37. Dua HS, Azuara-Blanco A (2000) Limbal stem cells of the corneal epithelium. Survey Ophthalmol 44:415–425
38. Ahmad S, Figueiredo F, Lako M (2006) Corneal epithelial stem cells: characterization, culture and transplantation. Regenerative Medicine 1:29–44
39. Nakamura T, Ang LPK, Rigby H, Sekiyama E, Inatomi T, Sotozono C et al (2006) The use of autologous serum in the development of corneal and oral epithelial equivalents in patients with Stevens-Johnson syndrome. Invest Ophthalmol Visual Sci 47:909–916
40. Chirila T, Barnard Z, Zainuddin, Harkin DG, Schwab IR, Hirst L (2008) Bombyx mori silk fibroin membranes as potential substrata for epithelial constructs used in the management of ocular surface disorders. Tissue Eng A 14:1203–1211
41. Hogan MJ, Alvarado JA, Weddell JE (1971) Histology of the human eye. Saunders, Philadelphia
42. Armitage WJ (1999) Eye banking. In: Easty DL, Sparrow JM (eds) Oxford textbook of ophthalmology, vol 2. Oxford University Press, Oxford, pp 1167–1171

Chapter 5
Banking of Heart Valves

Robert Parker

Introduction

The first aortic valve was implanted into a patient at Guys Hospital in London by Donald Ross in 1962 [1] and over the following 45 years major changes have occurred in the processing of heart valves. The first valves were preserved using chemical agents such as formalin, glutaraldehyde, beta propriolactone and in the latter years of the decade ethylene oxide. Prosthetic valves, mainly of the ball in the cage variety or tilting disc, were introduced in the previous decade to the allograft and in the sixties these were the two main types of valves that were used [2, 3]. Xenograft valves were introduced in the seventies being mainly the porcine aortic valve [4], with valves made from bovine pericardium following later in the decade [5]. Also around this time surgeons tried to manufacture heart valves from human tissue with autologous fascia lata [6] being the first material tried and later homologous dura mater [7], and both autologous and homologous pericardium being tried [8]. In the present millennium there are only really three types of valves regularly being implanted and these are bileaflet prosthetic valves, porcine xenografts (which can be stented or unstented) and allografts. Seventy five percent of the allografts are used nowadays in paediatric cases with a further 15% in adult congenital cardiac surgery and the final 10% in adult acquired surgery, particularly in redo operations and patients with bacterial or fungal endocarditis. The advantages and disadvantages of allografts are given in Table 5.1.

Selection Criteria

The donor criteria operated by the majority of heart valve banks allows age from new born to 60 years of age with no evidence of the standard virological markers such as HIV, Hepatitis B, Hepatitis C, HTLV or Treponema (syphilis). A previous

R. Parker (✉)
Heart Valve Bank, Brompton Hospital, Sydney Street, London SW3 6NP, UK
e-mail: R.Parker@rbh.nthames.nhs.uk

G. Galea (ed.), *Essentials of Tissue Banking*, DOI 10.1007/978-90-481-9142-0_5,

Table 5.1 Advantages and disadvantages of allografts

Advantages	Disadvantages
Blood flow through the valve orifice remains normal and anatomically correct (unstented xenografts give normal orifice but being from another species are not totally anatomically correct)	Unpredictable supply due to donation rate
No anticoagulation required and therefore less clinic appointments (particularly important in females who may become pregnant)	Greater risk of cross infection because of antibiotic disinfection rather then terminal sterilsation
No mechanical damage to blood cells that can occur with prosthetic valves	Cardiopulmonary bypass time during implantation is 50% greater than for prosthetic and stented xenograft valves
Gradual wear out of valve which is symptomatic, whereas prosthetic valves can fail shut suddenly or can lose disc or leaflet	Surgical technique is more complex
Low evidence of post operative endocarditis	Possibility of atheroma, fibrosis and calcium in valve due to donor age
Adaptable for paediatric and endocarditis surgery	More complex storage procedure and preparation at time of operation
Complete range of sizes. Prosthetic valves cannot be made below 15 mm diameter	Shorter valve life than prosthetic valve, though better patient survival

medical history of diseases that affect the connective tissue of the valve cusps such as rheumatic fever are also contraindications, but cardiac conditions such as myocardial infarction and left ventricular hypertension are acceptable for donation. The removal of the heart needs to occur within 24 h of death (although a few banks will accept up to 48 h if the donor's cadaver has been refrigerated within 2 h of death) and dissection of the valves from the heart should occur as soon as possible after cardiectomy and within a maximum of 12 h. The conditions of the area where the heart is removed from the donor should be as sterile as possible, so operating theatres or dedicated removal suites are preferable to post mortem rooms, and sterile instruments and receptacles should be used in all cases. In Europe about 50% of the donations are retrieved from heart beating donors undergoing multi-organ retrieval where the heart is either unsuitable for heart transplantation or cannot be placed, with about 4% of the donations being from hearts removed from patients undergoing heart transplantation and the remainder from non-heart beating donors, where some post mortem room retrieval will occur.

Processing and Disinfection of Heart Valves

The only valves that are routinely dissected nowadays are the aortic and pulmonary valves. Photographs of dissected aortic and pulmonary allografts are shown in Figs. 5.1 and 5.2. Mitral allografts [9] were used for a time in the 1990s and early 2000s, but problems arose with the chordae and myocardial connections and many

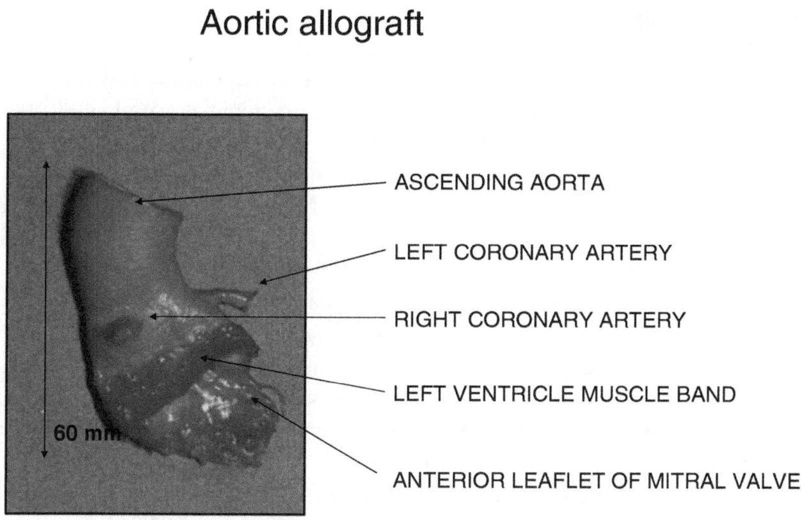

Fig. 5.1 Dissected aortic allograft

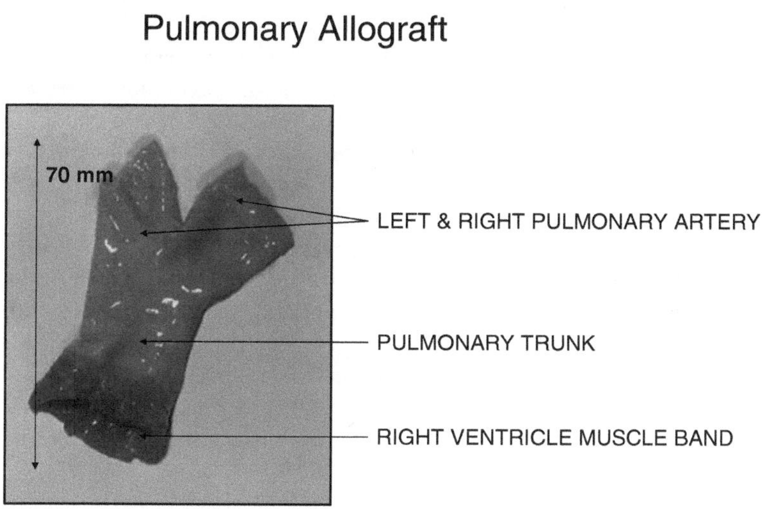

Fig. 5.2 Dissected pulmonary allograft

became incompetent, so very few centres use this type of valve nowadays. There were also surgical and technical difficulties as mitral valves do not form a ring in the same way that aortic and pulmonary valves do. Processing needs to be performed in a Class A environment and most banks perform the dissection in a cabinet at this level in a Grade B or C background. All heart valve banks worldwide use antibiotic disinfection methods using a combination of broad spectrum antibiotics with or without antifungal agents. Most antifungal drugs are cytotoxic so will affect the cell viability and therefore some banks who are concerned with this parameter now have omitted the antifungal agents, particularly amphotericin B [10–12], but this does mean that they will have a higher discard rate disinfection. The majority of banks in United Kingdom use a disinfection solution containing Gentamicin 4,000 mg/L, Vancomycin 50 mg/L, Polymixin B 200 mg/L, Imipenem 240 mg/L and Nystatin 50 mg/L, but the Nystatin has been replaced by Amphotericin B because of the non-availability of the former in sterile format in United Kingdom. The antibiotic cocktails have been formulated by microbiologists [13–15] to cover as wide a range of bacteria and fungi as possible and the levels of antibiotics in most is 20–50 times the minimum inhibitory concentration, but can be up to one thousand times. The anbiotic cocktails used by most banks give a disinfection rate of between 85 and 95%, meaning that about after immersion in the cocktail less than 15% of valves have viable bacteria or fungi rendering them unsuitable for clinical use. There is not a consensus of opinion at which is the most appropriate temperature to perform the antibiotic disinfection. The commonest temperature is +4°C, but a number of banks use room temperature (22–25°C) whereas in Australia 37°C is temperature most banks prefer. Germain at European Association of Tissue Banks Meeting in 2008 propounded the idea that the antibiotics are designed to be used at body temperatures and that this was the preferable temperature. Macdonald has presented work at conferences using similar assays on the antibiotic solutions used at heart valve banks in United Kingdom and showed that all were most effective at 37°C. The converse to this argument is that lysosomal activity is higher at this temperature and therefore more tissue degradation can be expected. Leeming [16] demonstrated that there is considerable carryover of antibiotics in the valve samples which could affect the accuracy of the microbiology testing and in the case of vancomycin in the leaflet this could equal the level in the antibiotic cocktail. Even after 20 washes the tissue samples show a high level of antibiotic residue. It is thought that the antibiotic residue could provide antibiotic prophylaxis to the patient as the endocarditis level following homograft implantation is lower than that for prosthetic and xenograft valves. However the residue could be a problem if the recipient of the valve is allergic to the antibiotic and there could be a small risk of systemic toxicity if the patient has hepatic or renal malfunction. It is probable that some of the antibiotics have a static effect rather than a cidal effect and it is possible that there could be resistant spores remaining which would explain why some heart valve cultures show bacterial or fungal contamination late during the culture.

The microbiology test performed by most heart valve banks involves putting samples of valve tissue (normally the aortic or pulmonary wall) into bottles of aerobic broth, anaerobic broth, sabourauds broth and incubating at 37°C and placing another sample in a general purpose broth and retaining at room temperature. Samples of the broth are sub-cultured on to respective agar plates at day six and the plates are read at day nine or day twelve. A fifth sample is used for testing for mycobacteria, which is commonly performed by heart valve banks but not always by the bankers of other tissues. There was evidence in the 1970s of transmission of TB with heart valve allografts and the advent of testing dates from this time [17]. In a recent paper Warwick and co-workers [18] have shown that in 38,413 donors in 24 heart valve banks over the last 15 years tested, none were positive for Mycobacterium tuberculosis, but 24 were positive for non-tuberculosis mycobacterium. The donor selection criteria for heart valves was not as strict in the 1970s and it is probable for this reason that no positive donors have been detected since. It has been questioned whether mycobacterial testing of heart valves should be continued, but it would seem important that because of the incidence of non tuberculous mycobacteria which have also been transmitted with porcine valves and also as a contaminant at time of cardiac surgery that the testing is continued. The commonest reason for non-tuberculous mycobacteria is water borne and is probably contaminating the tissue at time of removal in the post mortem room.

After antibiotic disinfection, valves are transferred to a isotonic solution containing a cryoprotective agent with dimethyl sulphoxide (DMSO) being the commonest in use. The solution would normally contain 10–15% of DMSO and a time of up to an hour is used for the valve to absorb the solution before cryopreservation. The valve is frozen at the rate of 1°C/min from +4°C to at least –80°C and is then stored either in ultra low temperature freezers at –140°C or the vapour phase of liquid nitrogen refrigerator at –175°C. Most banks have validated storage at this temperature for 5 years, but an increasing number are extending the validated period to 10 years and Mirabet has shown that it is possible to at least 13 years [19]. It is also important that the thawing of the valves is controlled as it has been shown by various groups that if valves are transferred directly from ultra low temperatures to body temperature water baths cracking can occur [20–22]. Most banks now instruct that there is an intermediate stage either by allowing the valve to thaw at the lower temperatures in the air before transfer to the waterbath or by placing the valve in solid carbon dioxide (cardice). The latter is the usual method of transfer from bank to operating theatre, so is being performed without realizing the significance of the stage.

Quality Assurance of Heart Valves

At the time of dissection valves are assessed for their quality. Many banks still use a fair, good, very good and excellent terminology, but the Heart Valve Bank in Rotterdam has tried to bring greater rationale to the system and have introduced a Quality Code List.

The list gives a Code to each valve depending on its abnormalities. The description of the Codes are as follows:

Acceptable for further processing[a]	
Code 01	No visible morphological abnormalities
Code 02	Minimal atheroma in basal attachment of leaflet
	Minimal fibrosis in (basal attachment of) leaflet/vascular wall
	Fenestration(s) in otherwise perfect graft
	Petechiae in otherwise perfect graft
Code 03	Minor atheroma in vascular wall (conduit)
	Atheroma in <1/3 of the basal attachment of the leaflet

	Fibrosis in <1/3 of (the basal attachment of) the leaflet
	Fenestrations
Code 04[b]	Atheroma in the vascular wall
	Atheroma in <2/3 of the basal attachment of the leaflet
	Fibrosis in <2/3 of (the basal attachment of) the leaflet
	Fenestrations
Code 05[b]	Discrete atheroma in the vascular wall or the basal attachmemt of the leaflet
	Discrete fibrosis in (basal attachment of) leaflets
	Pinpoint calcification in vascular wall
	Minor adhesions of leaflets at the commisures
	Fenestrations
Not acceptable for further processing	
Code 06	Extensive fibrosis or atheroma and/or calcification in vascular wall, calcification in leaflets
Code 07	Damaged during removal of the heart
Code 08	Damaged during dissection of the heart
Code 09	Incompetent valve
Code 10	Other abnormalities (anatomical or procedural)
Code 11	Failure to meet one or more of the time limits adherence requirements needed by bank.

Note: [a]Items are "and/or"
[b]Codes 04 and 05 are not acceptable for aortic valves from donors ≥ 56 years of age

It would be an excellent idea if banks who co-operated in the supply of heart valve allografts could adopt this nomenclature and could also adopt a standard method of sizing valves. At the present time there is no consistent measuring device for internal diameter of valves with some banks using obturators and other banks dilators and the position of the measurement with regard to cusp height is also variable. Likewise when it comes to length it is agreed that the starting point is at myocardium/wall junction but as aortic valve has curved arch and pulmonary artery bifurcates within 3 cm of valve, there is not a concensus opinion about the various lengths that need to be measured. As many surgeons obtain valves from more than one bank depending on supply it would be extremely helpful if all banks agreed a protocol for classification and sizing.

Current Issues in Heart Valve Banking

Viability and Storage

Throughout the history of heart valve banking, workers have tried to evaluate the best method of storage. There have been three main approaches to assess this and these can be summarized as viability, morphology and mechanical.

Most workers concentrated on viability in the early years of antibiotic disinfected valves with researchers looking at incorporation of tritiated thymidine into DNA, protein synthesis, lysosomal activity and tissue culture. The results from this determined the change from firstly dissolving the antibiotics in a balanced salt solution such as Hanks to secondly dissolving them in a nutrient solution like Medium 199 and in the last 20 years following this stage by cryopresevation. Al-Janabi and Ross [23, 24] using autoradiography showed that the tissue lost little viability during the time it was frozen at liquid nitrogen temperatures, so the main factor was how soon after dissection the valve was frozen and this conclusion was also reached by Gonzalez-Lavin [25] using a proline uptake method. Suh [26] showed that the length of warm ischaemic time was particularly important with a reduction from 92 to 56% between samples taken at 2 and 36 h post death. He also showed that a number of lysosomal enzyme levels were significantly increased during warm ischaemic times and concluded that valves should always have less than 12 h of warm ischaemia and no more than 24 h of cold ischaemia following this before cryopreservation. Brockbank [27] has shown that protein synthesis had a time dependent loss when valvular tissue was stored at +4°C or –80°C, with more rapid decline in the former, but this was not the case at liquid nitrogen temperatures.

Endothelial Cells

All of this research was looking at fibroblast rather than endothelial cell activity and the role of the endothelium has come more to the fore over the last 15 years. The relevance of the endothelial cells being viable has both positive aspects in that they protect the cusp from insudation of plasma components which are responsible for degeneration and calcification and negative aspects in that they are the most immunogenic of the cells in the heart valve and because of this can lead to rejection phenomena. Loose [28] has shown storage of valves at +4°C in nutrient medium leads to complete denudation of the endothelium within 21 days, but this was significantly decreased at –196°C. Work by Feng [29] confirmed that valves need to be stored below the glass point of water (–128°C) if endothelial viability is to be maintained. Yankah [30] from Berlin in 1987 showed infiltrates of lymphocytes on transplanted allogeneic valves and Yacoub [31] in same year demonstrated HLA Class I expression and Simon in 1998 Class II expression on surface of valve. In his 1993 paper Lupinetti [33]considered that the loss of endothelium may contribute to enhanced graft longevity by reducing the host immune response, which may contribute to valve degeneration, but on the other hand loss of endothelium may increase the capacity for thrombus foundation and adversely affect the underlying fibroblasts, thereby accelerating graft deterioration. Thus he concluded that it cannot yet be determined whether the loss of endothelial cells has a positive, negative or neutral effect on the long-term structure and function of valve allografts and 10 years on this is still the case.

Morphology of the Valves

In the present century there has been a switch from looking at viability of the tissue as it has never really been understood if viability was important to looking at the morphology of the valve after cryopreservation. Fischlein [34] showed that valves from heart beating donors had nearly intact endothelium if cryopreserved within 24 h but there was a lack of endothelial cells in non heart beating donors with oedema and vacuolization as well. Brockbank [35] has proposed that a possible cause of tissue deterioration and calcification could be interstitial ice formation causing damage to the extracellular matrix and is now suggesting that vitrification may be a better option. Villalba [36] has observed that in cryopreserved valves 4.7% of cells have irreversible changes and 23.5% show apoptotic signs, whereas valves before freezing only show reversible cell injury. He postulates that the apoptosis could be caused by hypoxia, ischaemia, free radicals of oxygen or calcium ionophores but does not consider that disinfection and dimethyl sulphoxide exposure alone are responsible for this, although they enhance the rate of cell death. Schenke-Layland [37] has demonstrated serious alteration and significant deterioration of collagenous and elastic fibre structures, accompanied by a general damage of the leaflet histoarchitecture caused by extracellular ice formation. It is considered that this type of damage can lead to calcification of the valve tissue and also without the elastic properties the valve could become incompetent.

Mechanical Properties of Valves

If viability is not important and the most important factor is that the allograft valve should mimic the action of the native valve, a number of workers have considered that the mechanical properties of the valve may be the best parameter to evaluate after preservation. Wright and Ng [38] devised a method in 1974 for measuring the elasticity of human aortic valve cusps and this was used by staff at National Heart Hospital to compare the elasticity of valves after storage in nutrient medium. The results of this was that there was a slow decline in elasticity and tensile strength over the first 3 weeks which became faster thereafter. Wassenaar [39] reported that cryopreservation resulted in a considerable reduction of contraction in response to potassium, but that the response to noradrenaline, endothelin-1 and prostaglandin F remained unchanged. Vesely [40] compared the mechanical behaviour of cryoporeserved allograft leaflet to that of fresh tissue and xenografts by measuring their bending stiffness and their uniaxial tensile stress/strain and stress/relaxation behaviour. The results for the bending tests showed no significant difference between the pliability of cryopreserved allografts and fresh tissue but the xenograft material which had been treated with glutaraldehyde was significantly stiffer than both. The transition from a low to a high modulus on the stress/strain curves, a measure of extensibility was also similar for fresh and cryopreserved tissue but less for xenograft.

It can therefore be concluded that most cryopreservation methods are satisfactory and retain the physical structure and biochemical profile of the valve at similar levels to that at the time of freezing. It may therefore be important as Gall and O'Brien [41] have shown that it is important that heart valves are procured as soon as possible after cessation of heart beat and should be cryopreserved within 1 or 2 days if not clinically implanted. Yacoub [42] has shown that his homovital valves, which came either from heart transplant recipients or multi organ donors where the heart could not be used for transplantation and had very low level antibiotic disinfection and were implanted within 24 h produced better results than the cryopreserved valves.

Conclusions

Since the small beginnings in 1962 the number of heart valve banks worldwide has now increased in 2007 to 71 and these banks dissect aortic and pulmonary valves from over 11,000 hearts per year. Sixty six percent of the activity is in North America and 26% in Europe. In the United States, most of the heart valve banking is performed by two large companies whilst in all other countries the banking is performed either by hospitals or the national blood service of the country. In the early years of heart valve banking, only the aortic valve was collected but since the 1980s the demand for the pulmonary valve has increased and it is now the more requested valve, see Fig. 5.3. The main reason for this is that pulmonary allografts are more successful in congenital cases than prosthetic and xenograft valves, both in children and adults, whilst the difference is not as great in the aortic position, where valve replacement is mostly for acquired conditions. Allografts do not need anticoagulation so will also be the valve of choice in women of child bearing age who could become pregnant and children who because of the long term side effects should not

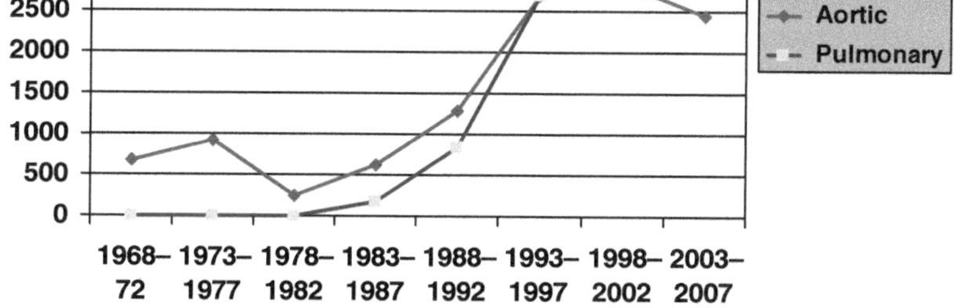

Fig. 5.3 Number of aortic and pulmonary valve allografts used in 5 year cohorts from 1968–2007

have anticoagulation therapy for their entire lives. The allograft gives greater flexibility in insertion as it does not have a fixed sewing ring so allows the surgeon to shape the valve for complex congenital conditions or to replace diseased tissue in endocarditis cases. The non-stented xenograft which has been introduced over the last 10 years is also showing good results in these operations as well.

Allografts in Europe, North America and Australia account for about 5% of the aortic valve replacements performed in these countries and 90% of the pulmonary valve replacements. The total market share for allografts in United States in 2007 was 4% with prosthetic valves accounting for 20% and xenograft valves for 76%.

In the future it is likely that decellularised allografts will be used as this could remove any problems with immunology and methods have been postulated for this in several countries. Cryolife in United States have now marketed this type of valve for 3 years and Francisco da Costa [43] in Brazil has shown that over a 4 year period there has been a 100% patient survival with decellularised valves compared to 91% for cryopreserved valves. Most involve the use of either deoxycholic acid or sodium dodecylsulfate. Research is also looking at seeding these valves with fibroblasts or endothelial cells grown from recipient tissue which could reduce thrombolytic complications. The use of biological scaffolds is also being investigated which would also need to be seeded with recipient or possibly foetal stem cell derived tissue.

References

1. Ross DN (1962) Homograft replacement of the aortic valve. Lancet 2/7254:487
2. Starr A, Herr RH, Wood JA (1965) The present status of valve replacement. Acta Chirurgia Scand 374:1–87
3. Bjork VO, Holmgren A, Olin C, Ovenfors CO (1971) Clinical and haemodynamic results of aortic valve replacement with Bjork-Shiley tilting disc valve prosthesis. Scand J Thorac Cardiovasc Surg 5:177–191
4. Carpentier A (1971) The concept of biorposthesis. Thoraxchirurgie vaskulare chirurgie 19:379–383
5. Mary DA, Pakrashi BC, Catchpole DW, Ionescu MI (1975) Tissue valves in the mitral position: 5 years experience. Br Heart J 37:1123–1132
6. Ionescu MI, Ross DN, Deac R et al (1970) Autologous fascia lata for heart valve replacement. Thorax 25:46–56
7. Zerbini EJ (1975) Results of replacement of cardiac valves by homologous dura mater valves. Chest 67:706–710
8. Deac RF, Simionescu D, Deac D (1995) New evolution in mitral physiology and surgery: mitral stentless pericardial valve. Ann Thorac Surg 60:5433–5438
9. Acar C, Tolan M, Berrebi A et al (1996) Homograft replacement of the mitral valve selection, technique of implantation and results in 43 patients. J Thorac Cardiovasc Surg 111:367–380
10. Agvirregoicoa V, Kearney JN, Davies GA, Gowland G (1989) Effects of antifungals on viability of heart valve cusp derived fibroblasts. Cardiovasc Res 23:1058–1061
11. Brockbank KG, Dawson PE (1993) Cytotoxicity of amhotericin B for fibroblasts in human heart valve leaflets. Cryobiology 30:19–24
12. Birtsas V, Armitage WJ (2005) Heart valve cryopreservation: Protocol for addition of dimethyl sulphoxide and amelioration of putative amphotericin B toxicity. Cryobiology 50:139–143
13. Waterworth PM, Lockey E, Berry EM, Pearce HM (1974) A critical investigation into the antibiotic sterilization of heart valve homografts. Thorax 29:432–436

14. Wain WH, Pearce HM, Riddell RW, Ross DN (1977) A re-evaluation of antibiotic sterilization of heart valve allografts. Thorax 32:740–742
15. Yacoub M, Kittle CF (1970) Sterilization of valve homografts by antibiotic solutions. Circulation 41(Suppl II):29–32
16. LeemingJP, Lovering AM, Hunt CJ (2005) Residual antibiotics in allograft heart valve tissue samples following antibiotic disinfection. J Hosp Infect 60:231–234
17. Anyanwu CH, Nassau E, Yacoub M (1976) Miliary tuberculosis following homograft valve replacement. Thorax 31:101–106
18. Warwick RM, Magee JG, Leeming JP et al (2008) Mycobacteria and allograft heart valve banking: An international survey. J Hosp Infect 68:255–261
19. Mirabet V, Carda C, Solves P et al (2008) Long term storage in liquid nitrogen does not affect cell viability in cardiac valve allografts. Cryobiology 57:113–121
20. Hunt CJ, Song YC, Bateson EA, Pegg DE (1994) Fractures in cryopreserved arteries. Cryobiology 31:506–515
21. Pegg DE, Wusteman MC, Boylan S (1997) Fractures in cryopreserved elastic arteries. Cryobiology 34:183–92
22. Wassenaar C, Wijsmuller EG, Van Herwerden LA et al (1995) Cracks in cryopreseved aortic allografts and rapid thawing. Ann Thorac Surg 60:S165–S167
23. Lockey E, Al-Janabi N, Gonzalez-Lavin L, Ross DN (1972) A method of sterilizing and preserving fresh allograft heart valves. Thorax 27:398–400
24. Al-Janabi N, Ross DN (1974) Long term preservation of fresh viable aortic valve homografts by freezing. Br J Surg 61:229–232
25. Gonzalez-Lavin L, McGrath LB, Amini S, Graf D (1987) Determining viability of fresh and cryopreserved homograft valves at implantation. Heart Vessels 3:205–208
26. Suh H, Lee JE, Park JC et al (1999) Viability and enzymatic activity of cryopreserved heart valves. Yonsei Med J 40:184–190
27. Brockbank KG, Carpenter JF, Dawson PE (1992) Effects of storage temperature on viable bioprothetic heart valves. Cryobiology 29:537–542
28. Loose R, Markant H, Sievers H, Bernhard A (1993) Fate of endothelial cells during transport, cryopresevation and thawing of heart valve allografts. Transplant Proc 25:3247–3250
29. Feng XJ, Van Hove CE, Walter PJ, Herman AG (1996) Effects of storage temperature and fetal calf serum on the endothelium of porcine aortic valves. J Thorac Cardiovasc Surg 111: 218–230
30. Yankah Ac, Wessel U, Dreyer H et al (1987) Transplantation of aortic and pulmonary allografts, enhanced viability of endothelial cells by cryopreservation, importance of histocompatability. J Cardiac Surg 2:S209–S220
31. Yacoub MH, Festenstein H, Doyle P et al (1987) The influence of HLA matching in cardiac allograft recipients receiving cyclosporine and azothioprine. Transplant Proc 19:2487–2489
32. Simon A, Wilhelmi M, Steinhoff G et al (1998) Cardiac valve endothelial cells; relevance in the long term function of biological valve prosthesis. J Thorac Cardiovasc Surg 116:609–616
33. Lupinetti FM, Tsai TT, Kneebone JM, Bove EL (1993) Effect of cryopreservation on the presence of endothelial cells on human vein allografts. J Thorac Cardiovasc Surg 106:912–917
34. Fischlein T, Schutz A, Uhlig A et al (1994) Integrity and viability of homograft valves. Eur J Cardiothorac Surg 8:425–430
35. Brockbank KM, Lightfoot FG, Song YC, Taylor MJ (2000) Interstitial ice formation in cryopreserved homografts:a possible cause of tissue deterioration and calcification in vivo. J Heart Valve Dis 9:200–206
36. Villalba R, Pena J, Luque E, Villagran JLG (2001) Characterization of ultrastructural damage of valves cryopreserved under standard conditions. Cryobiology 43:81–84
37. Schenke-Layland K, Madershahian N, Riemann I et al (2006) Impact of cryopreservation on extracellular matrix structures of heart valve leaflets. Ann Thorac Surg 81:918–927
38. Wright JEC, Ng YL (1974) Elasticity of human aortic valve cusps. Cardiovasc Res 8:384–390

39. Wassenaar C, Bax WA, Van Suylen RJ et al (1997) Effects of cryopreservation on contractile properties of porcine isolated aortic valve leaflets and aortic wall. J Thorac Cardiovasc Surg 113:165–172
40. Vesely I, Gonzalez-Lavin L, Graf D, Bouchner D (1990) Mechanical testing of cryopreserved aortic allografts, comparison with xenografts and fresh tissue. J Thorac Cardiovasc Surg 99:119–123
41. Gall K, Smith SE, Willmette CA, O'Brien M (1998) Allograft heart valve viability and valve processing variables. Ann Thorac Surg 65:1032–1038
42. Yacoub M, Rasmi NR, Sundt TM et al (1995) Fourteen year experience with homovital homografts for aortic valve replacement. J Thorac Cardiovasc Surg 110:186–194
43. da Costa FD, Santos LR, Collatusso C et al (2009) Thirteen years experience with the Ross operation. J Heart Valve Dis 18:84–94

Chapter 6
Banking of Skin

Ellen Heck

While early excision of the burn wound certainly proceeds the establishment of widely available allograft skin banks, this form of burn treatment was greatly enhanced by the 1970s proliferation of skin banks. Prior to the establishment of skin banks, commonly in conjunction with burn centers, physicians relied on the random or chance availability of allograft skin to supplement limited autografts. However, as skin banks flourished, escher removal by early excision became not only desirable, but the standard of care.

Skin banking across the world adopted various methods of storage including fresh, frozen, glycerol preserved and irradiated. The questions most frequently debated about the desirability of methodologies centered on cell viability and structural integrity. These same debates continue, especially as they relate to viability [1]. However, there is more consensus on the needs for structural integrity. To perform as desired as an extended, but non permanent barrier, the allograft skin must be comprised of the layers seen in Fig. 6.1.

The stratum corneum of the epidermis creates the barrier function while dermis allows for in growth of tiny blood vessels which provide nourishment to attenuate the grafts stability. Storage methods such as fresh and frozen skin allografts which preserve the layers in as near normal state as possible are preferred by many physicians [2]. Some research has also been done on preservation at room temperature or 22°C incubation for graft storage in nutrient media [3]. Grafts which are glycerol or irradiation preserved are questioned as to collagen alteration, viability, and other functional qualities. However, for ease of storage, they may be most desirable in some settings where frozen storage or even refrigerated storage is problematic (Fig. 6.2). Frozen allografts are subject to viability and structural changes not just during cryopreservation but also during thawing. The most universally accepted freeze-thaw protocols rely on slow freezing and rapid thawing. The intent is to minimize any cell membrane damage due to fluid transfer. Process of glycerolysation

E. Heck (✉)
Transplant Medical Services, UT Southwestern Medical Center, Dallas, TX 75390-9074, USA
e-mail: Ellen.Heck@UTSouthwestern.edu

Preparation assistance by Marilyn Hayes and William Timmons, RN

G. Galea (ed.), *Essentials of Tissue Banking*, DOI 10.1007/978-90-481-9142-0_6,

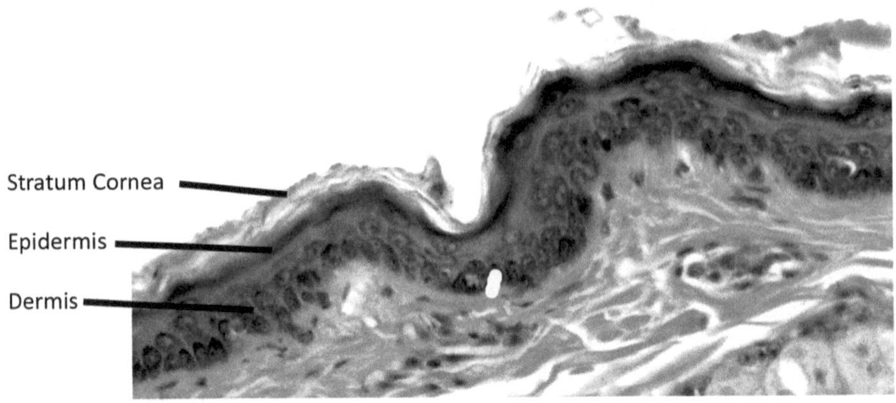

Fig. 6.1 Skin layers

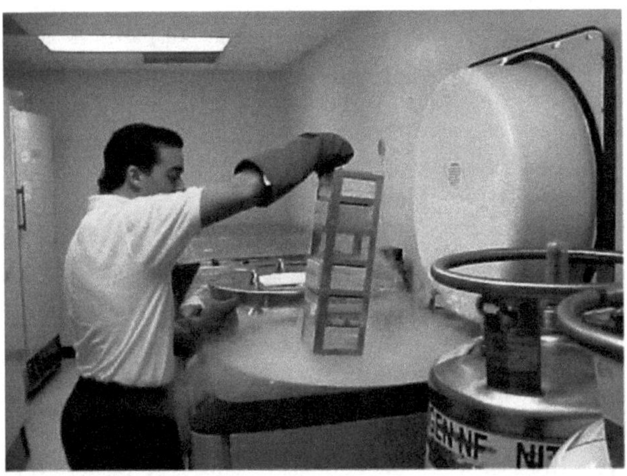

Fig. 6.2 Frozen skin allograft storage

or irradiation may also compromise cellular integrity and reconstitution methods should be controlled to adequately rinse and rehydrate to achieve the desired graft adherence parameters desired. No currently available allograft meets all the functions of autograft. Which allograft is most desirable depends on the given situation and which factors can be best achieved by a specially preserved graft. The factors to consider are availability, length of anticipated need for coverage, and of course accessibility to appropriate storage conditions.

The Dutch Burn Foundation Skin Bank utilizes glycerol preserved skin allografts preferentially and reports both success and satisfaction with this method [4].

However, before we discuss procurement, processing and storage in more detail, we should give some consideration to donor acquisition. How do we obtain donors and briefly how are they screened and entered into the donation process.

Acquisition

A robust awareness and education program contributes to sustaining donation programs whether this is by first person or by next of kin consent. In skin banks providing fresh allografts, active donation programs can have a major impact. Public and professional awareness of the needs and benefits of tissue donation is an important component in the success of a skin bank. The success stories are compelling and proper education to dispel misinformation is important. Skin bank public and professional education programs may be independent or part of a joint efforts with other entities.

Screening and testing also impacts graft availability and once again this is particularly true when trying to provide fresh allografts. Screening questions are most often completed prior to procurement but serology testing and microbiology testing can significantly delay tissue release. Serology testing, antigen, antibody and nucleic acid tests (NAT) usually can be completed within 24 h of receipt by the reference laboratory of the appropriate donor blood samples. All testing regimes should comply with government regulations and peer review standards. Unlike serologies, microbiology will usually require a minimum of 48 h and preferably 72 h for fresh skin release. Seven to 14 days is the customary microbiologic incubation time for long term graft storage. Laboratories need to be informed of the concerns the skin bank and burn center may have for making fresh grafts available so that the maximum information can be obtained in the minimum amount of time. Doing gram stains and sub cultures more frequently than may be "routine" for a reference laboratory can be the difference in having enough information available to release fresh grafts [5]. While some bacteria are normal flora and not a concern in grafting, other organisms, pathogens, or opportunistic pathogens can cause infection or life threatening sepsis in burn patients [6]. Of course having more allograft donors which may be held for possible use as fresh will mitigate some of the concern for rapid response on procurement cultures.

Retrieval

Donor allograft skin may be procured up to 12 h without prior refrigeration of the body and up to 18 h for first incision with prior cooling/refrigeration [7]. However, prior circumstances may impact these parameters especially when death occurs where extreme temperatures may play a role in body decomposition. Once verification has occurred of all required information, consent, time of death, refrigeration and available medical social history, the actual process of allograft recovery can begin.

Supplies and instrumentation may vary from program to program depending on each bank's policy and procedures but will include some basic similarities as indicated in Table 6.1.

Table 6.1 Retrieval supplies

Prep supplies
1. Prepared "Prep Kit" or razor and sponge
2. Basin
3. Antiseptic agent – iodophor
4. Alcohol
5. Gauze sponges (4 × 4) or similar

Instruments[a]
1. Dermatome
2. Dermatome blades
3. Forceps
4. Scissors

Draping and sterile field supplies[a]
1. Drape sheets
2. Caps
3. Gowns
4. Gloves
5. Masks
6. Scrub brushes
7. Sterile basin
8. Sterile bottles/transport containers

[a]Instruments and sterile supplies should be packaged sterile in appropriate wraps and/or peel packs to permit delivery to the sterile field without contamination.

Instruments and supplies must have dating appropriate to the method of sterilization and a supply log such as seen in Table 6.2 should be maintained in the donor record for each retrieval.

Personnel involved in retrieval of tissue should be thoroughly trained in the site assessment and preparation, sterile techniques and surgical techniques required for graft removal. If only allograft skin recovery is to occur a team of two trained individuals should be sufficient to accomplish the preparatory work and the recovery and circulating assistant functions.

Prior to beginning retrieval, the technician should perform a detailed physical inspection. This inspection should concentrate on signs of high risk behavior such as needle tracts and certain types of tattoos which could be indications of injection of illicit drugs (Fig. 6.3). Recent tattoos and piercing where shared needle and ink might have been used are also of concern for disease transmission.

Following a standard assessment protocol with an approved policy and procedure and implement form will aid in this process (Fig. 6.4).

Donors discovered to have questionable observations on physical assessment should be reviewed with available medical consults, re-checked with donor history and/or deferred.

Surgical preparation of the donor consists of a general cleansing to remove debris and/or environmental contaminants followed by hair removal. Hair removal should be carefully performed to prevent nicks and abrasions to the epidermal skin layer.

Table 6.2 Recovery supply log

TRANSPLANT SERVICES CENTER SOUTHWESTERN	**Tissue Recovery Supplies Log**	**TSC Donor ID:**__________ Cross-Ref. ID(s):

Supply Checklist

Sterile Supplies:				***Solutions / Fluids:***		
Bone Recovery Packs (MTF)	Set of 3	☐		Povidone-Iodine Scrub (500 mL)	1	☐
Bone/Vein Recovery Instruments	2	☐		Povidone-Iodine Solution (500mL)	1	☐
Air / Electric Dermatome	1	☐		Hibiclens (500 mL)	1	☐
Dermatome Blades (Zimmer)	5	☐		10% Bleach (500 mL)	1	☐
Skin bottles	4	☐		Plasmalyte-A (500 mL)	3	☐
Skin basins	4	☐		Lactated Ringer's (500 mL)	2	☐
Pipettes	4	☐		Isopropyl Alcohol Pints	4	☐
Culturettes	26	☐		Liquid Soap	1	☐
Scrub brushes (Hibiclens, Betadine)	16	☐		Disinfectant (Spray bottle)	1	☐
Dura-Prep, Large	5	☐		***Protective Apparel:***		
Drapes - medium	5	☐		Caps / Bonnets	6 each	☐
Impervious drape sheets	4	☐		Masks / Masks with shields	6 each	☐
Steri drapes	2	☐		Shoe Covers / Boots	6 pair	☐
Sterile Gowns	3	☐		Disposable Aprons / Gowns	6 each	☐
Gauze 4x4's (Large)	3	☐		***Red tackle box containing:***		
Bag Spikes	2	☐		50% Dextrose	2	☐
Scalpel Blades	10	☐		Gentamycin	2	☐
Gigli saw blades	4	☐		18 Gauge Spinal Needles	2	☐
Sterile scissors	1	☐		18 Gauge Needles	8	☐
Non-sterile Supplies:				35cc Syringes	3	☐
DDI Kit (MTF blood draw supplies)	1	☐		12cc Syringes	4	☐
Spool Suture Line	1	☐		Dura-Prep (Small)	4	☐
Prosthetics (lowers & uppers)	4	☐		Alcohol Swabs	12	☐
Disposable Razors	6	☐		TSC Blood Pack	1	☐
Disposable Head Block	1	☐		Extra Blood Tubes (Red, Purple)	4	☐
Union-all / Body Bag (1 of each)	2	☐		Sharps Container (Small)	1	☐
Large Trash Bags	4	☐		Extra Culturettes	6	☐
Biohazard Bags	5	☐		Calculator	1	☐
Non-sterile sheets	6	☐		Timer	1	☐

Supplies to be added prior to departure for recovery:			If Needed:
☐ Donor Chart	☐ Sterile gloves	☐ Polyfoam packer / wet ice	☐ Ocular recovery kit
☐ Camera	☐ Cut-resistant gloves	☐ Dermatome motor	☐ Vein recovery supplies
☐ Culture tubes / labels	☐ Prep gloves	☐ __________	☐ Heart recovery supplies

TRAVEL KITS CONTAIN ITEMS NEEDED FOR SKIN AND BONE RECOVERY, BRING ADDITIONAL SUPPLIES AS NEEDED

Form #: F-RP-501	Revision Date(s): 5/25/05, 3/2006, 9/12/06	Page 1 of 1

Hair removal should be accomplished in all areas where grafts will be recovered and in all areas immediately adjacent to these areas.

Once hair has been removed, the donor body is prepped with a surgical scrub; e.g iodophor, chlorhexidine gluconate. Contact of the scrub solution with the skin should follow manufacturer's recommendations usually for a minimum of 3–10 min

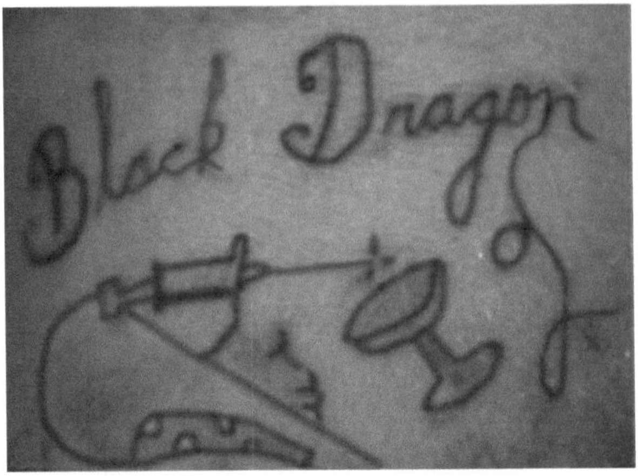

Fig. 6.3 Drug related tattoo

duration. The prep solutions should then be removed with an alcohol or similar rinse which will assist with cleansing along with removal of prep product residue which might interfere with reliable recovery skin cultures for bacteria and fungus.

The next step is the establishment of a sterile field. Draping should occur of all areas not involved in the surgical removal of tissue. Sterile disposable drape sheets or other operating room appropriate draping material is used for this procedure, see Fig. 6.5.

A sterile back table or work area should also be established for the placement of instruments and sterile supplies used in the recovery. Technicians should wear caps, masks, gowns and gloves during the recovery procedures. The technicians should also perform a surgical hand scrub after donning cap and mask and prior to gowning and gloving. Operating room nurses and/or technicians are often good training resources for aseptic technique as well as operating room manuals.

Skin is generally removed from the legs and back, but may be taken from any area consistent with the consent specifications and technical limitations. Consideration should be given to funeral arrangements and embalming concerns of funeral directors as fluid leakage may occur post graft removal. Providing a plastic garment [8] to the funeral home may be helpful in preventing leakage and in improving relationship between donations programs and funeral homes. Skin thickness may range from 0.012 to 0.018 inch with surgeon preference as an additional consideration.

Allograft skin utilized for primary burn wound excision is split thickness skin [9] with both epidermal and dermal layers. The epidermal layer provides the barrier function to reduce evaporative fluid loss and aid in the control of bacterial colonization. This additionally helps with pain reduction. The dermal layer allows for the nutrients from the excised wound bed to prolong the stability and adherence of the graft by ingrowth in to the allograft. Graft adherence is dependent in part on a uniform dermal bed in the allograft to maintain the adherence to the epidermal layer and to the excised wound bed. Dermatome setting and graft strips should be checked frequently to assure settings are providing desired consistency. Dermatome blade

TRANSPLANT SERVICES CENTER SOUTHWESTERN	**Physical Assessment Record**	**TSC Donor ID:**____________ Cross-Ref. ID(s):

Recovery Team Assessment (continued)

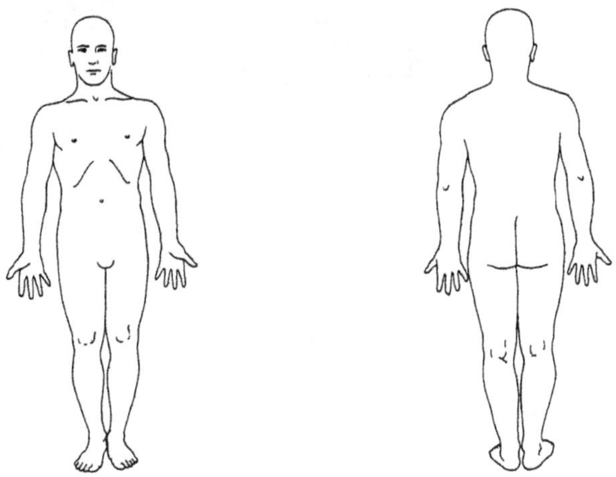

SUPINE
□ Unremarkable – No findings

PRONE
□ Unremarkable – No findings

Use the following codes to illustrate the location and type of physical characteristic noted during the physical assessment. Physical findings requiring description should be documented in the Comments Section.

Codes for physical findings:

1) Abrasion	9) ID Band / Tag	17) Scar (Surgical)
2) Autopsy incision	10) IV / Arterial line	18) Scar (Traumatic)
3) Bruise / Contusion	11) Jewelry (describe)	19) Skin Lesion / Rash (describe)
4) Body Piercing	12) Lacerations	20) Tattoo (describe)
5) Cast / Orthopedic device	13) Mole (describe)	21) Urethral catheter
6) Deformity	14) Needle entry site (therapeutic)	22) Other:________
7) Dressing / Bandage	15) Organ recovery incision	23) Other:________
8) ET tube / NG tube	16) Surgical incision	24) Other:________

Comments:____

Donor determined to be acceptable for recovery following completion of physical assessment: □ **Yes** □ **No**

Signature of person responsible for form completion	Title	Date

Form #: F-SP-525	Revision Date(s): 3/18/08	Page 2 of 3

Fig. 6.4 Physical assessment form

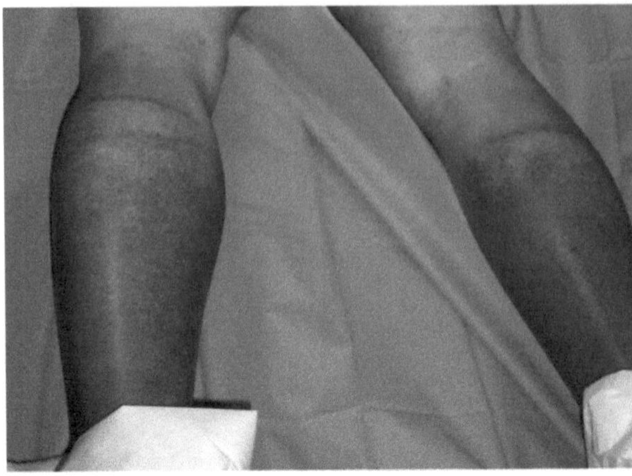

Fig. 6.5 Sterile field

settings may require adjustments for varying body surface areas. Graft thickness is a function of the following:

1. Dermatome or blade setting
2. Angle to plane surface
3. Downward pressure
4. Advancement speed
5. Body area
6. Donor age and skin condition
7. Operator skill
8. Postmortem time and death related factors

Figure 6.6 shows the positioning of dermatome relative to the plane surface and the application of drive and pressure. The dermatome needs a slight downward pressure while activating the air or electric control mechanism and a steady forward movement. To finish a cut, tilt or angle the blade upwards and release the cutting

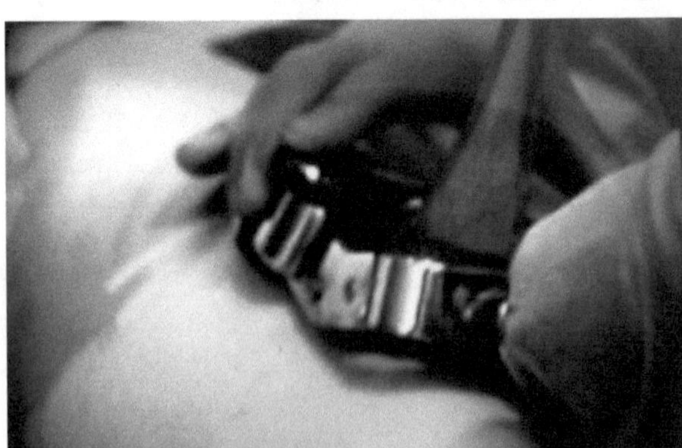

Fig. 6.6 Dermatome skin recover

control. Blades should be changed between each independent recovery site and whenever necessary due to dulling. Posterior and anterior sites require re-prepping, draping and aseptic precautions [7]. Positioning of the dermatome in such a way as to appear to slightly overlap one strip with another will assist in compensating for the area between the head width and actual blade cutting area. This will help to increase yield from all dermatomed areas.

Preservation and Storage

As eluded to earlier, methods for preservation may vary based on resources available and surgeons preferences. Frozen allografts are widely used in many parts of the world. Allograft skin processing for freezing may follow the following steps:

1. Placement in media for transport
   a. Balance nutrient media usually with addition of antibiotic
   b. Antibiotic using broad spectrum antibiotic; e.g., gentamycin, kanamycin, penicillin
2. Place transport container on ice and preserve cooling until refrigeration
3. Refrigerate 2–8°C.
4. Perform pre-freezing steps
   a. Add cryoprotective media; i.e., glycerol or DMSO.
   b. Return to refrigeration for approximately 30–60 min to allow for penetration of the protectant solution into skin layers
   c. Under a laminar flow hood or filtration controlled environment, while retaining chilling, place graft on backing material for freezing.
   d. Grafts should be trimmed as necessary to achieve as uniform pieces as possible
   e. Place prepared grafts in double bag system and label as appropriate to size and number of pieces.
   f. Each bag should contain a unique donor and piece number to facilitate tracking from donor to recipient; i.e., 2009-1138 1–4 12 inch
   g. Completed packages are the frozen by control rate freezing or heat sink at –1 to –4°C temperature drop per minute (Fig. 6.7)
   h. Freezing method should compensate for exothermic heat plateaus.
   i. Frozen grafts may be maintained for up to 5 years in temperatures from –70 to –196°C depending on the packaging.

Once frozen, grafts should be transported and maintained in their frozen state until readied for application. Grafts may be transported on dry ice or other transport system which will maintain a −45 to −70°C environment. Thawing and refreezing is not recommended due to possible structural and viability compromise.

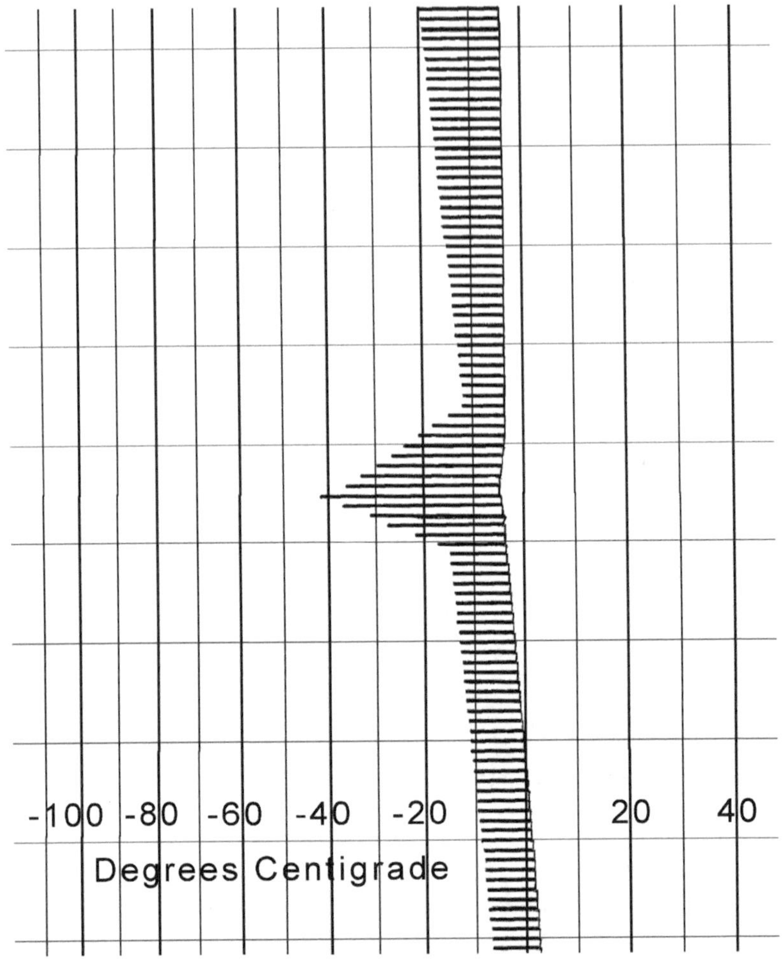

Fig. 6.7 Graph showing controlled rate freezing

Thawing of grafts for application should be accomplished rapidly at temperatures between 25 and 35°C. Cryoprotectants should be thoroughly rinsed from the graft as it may be toxic to the allograft skin at room temperature or above. Also, it may interfere with graft adherence and cause pain and burning sensation or reaction in some patients. All thawed grafts are considered for single application only as they are usually thawed in the operating suite while patient procedures are being performed. Thawed grafts, if properly stored in nutrient tissue culture media, may be maintained in refrigerator conditions for later application to the same patient. These grafts may be stored up to 4 days without media changes although media changes at 72 h may be more desirable if viability is of concern.

Another graft preservation method utilized in many countries is glycerol preservation. In this method, the allograft skin is treated with varying concentrations of

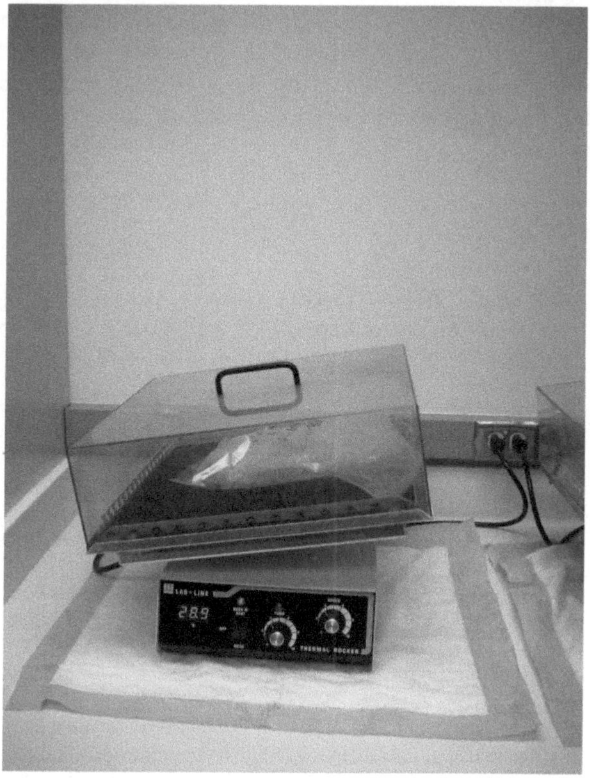

Fig. 6.8 Second stage of glycerol preservation

glycerol until it reaches its final state of dehydration/preservation. The Dutch Burn Foundation Euro Skin Bank routine utilizes this method. The grafts are treated within 72 h of retrieval in 50% glycerol with progression through several changes of various concentrations of glycerol before finally reaching a final 85% storage concentration [4] (Fig. 6.8). Glycerol preserved skin may also be stored up to 5 years and is stored at ambient room temperatures of 22°C (+ –) or at refrigeration of 2–8°C.

Conclusions

The skin adherence and barrier function, regardless of the preservation method employed, may be difficult to evaluate. The nature of the burn wound poses many challenges to the allograft performance. The appropriateness of excision depth, fluid accumulation, wound colonization and location of wound site all contribute significantly to the performance of the allograft. Nevertheless, allograft skin remains a consistent preference in burn wound therapy. As a preparation and test for autografting and as a functional cover to augment limited autografts in major burn injury, physicians choose allografts. These allografts need to meet size and structure requirements and provide ease of application to reduce operating theatre time and provide application with the minimum number of seams to reduce bacterial access/egress. A well prepared functional allograft is an asset to burn treatment and recovery.

References

1. Koizumi T, Robb EC (2006) Plessinger RT, Kagan RJ Abstract: prolongation of cadaveric skin viability with polyphenol: 174. J Burn Care Res. 38th Annual Meeting, Tuesday Through Friday, April 4–7, 2006, Caesars Palace, Las Vegas, Nevada. 27(2) Supplement:S136, Mar 2006
2. Gomez M, Cartotto R, Knighton J, Smith K, Fish JS (2008) Improved survival following thermal injury in adult patients treated at a regional burn center. J Burn Care Res 29(1): 130–137, Jan/Feb 2008
3. Robb EC, Bechmann N, Plessinger RT, Boyce ST, Warden GD, Kagan RJ (2001) Storage media and temperature maintain normal anatomy of cadaveric human skin for transplantation to full-thickness skin wounds. J Burn Care Rehabil 22(6):393–396, Nov–Dec 2001
4. Euro Skin Bank. http://www.euroskinbank.nl/
5. Neely AN, Plessinger RT, Stamper B, Kagan RJ (2008) Can contamination of a patient's allograft be traced back to the allograft donor? J Burn Care Res 29(1):73–76, Jan/Feb 2008
6. Heck, Ellen BS, MT; Blood, Sharron BS, MT; Baxter, Charles MD (1981) The importance of the bacterial flora in cadaver homograft donor skin: bacterial flora in cadaver homograft. J Burn Care Res 2(4):212–215
7. American Association of Tissue Banks (2008) Standards for Tissue Banking, 12th edn. AATB
8. Pierce Chemical Company. http://www.piercechemical.com
9. Heck E (1987) Skin banking: a basic overview. Tissue banking. American Association of Blood Banks

Part II
Principles of

Chapter 7
Storage, Processing and Preservation

John N. Kearney

Introduction

Tissue Replacement

Tissues can malfunction as a result of numerous processes including congenital malformation, disease and damage. Most tissues in the human body have a limited capacity for regeneration, in response to such malfunction.

Skin, being in direct contact with the environment does have regenerative capacity in order to deal with the minor abrasions and lacerations of everyday life. In fact the outer layer, the epidermis, is replaced continuously from below, and the outer skin scales (squames) are shed into the environment. However, the capacity to regenerate skin declines as the depth of the wound increases. Above a certain critical size, full thickness skin wounds fail to regenerate and in this case the only way to achieve functional skin is to use a skin graft, either an autograft or allograft.

Bone is another tissue which turns over continuously and hence can repair fractures but removal of a large section of bone resulting from traumatic damage or eg tumour excision, will not result in regeneration or replacement. In fact removal of a piece of bone above a critical size results in a permanent defect. It appears that the healing response is designed to resorb and replace dead bone rather than to regenerate bone that is missing. Therefore this leads to one of the major strategies in bone tissue transplantation i.e. to replace missing bone with either autologous or donor bone to stimulate the resorption and replacement of the graft with new autologous bone.

Other tissues do not turnover continuously e.g. cartilage. Once formed, there is little turnover of the extracellular matrix and the cells are largely quiescent through life. Therefore in this case there is little that can be done to stimulate a natural tissue regeneration response. Here the only hope is to replace defective cartilage with a fully viable allograft, or to stimulate the generation of new autologous cartilage in vitro using autologous chondrocytes [1].

J.N. Kearney (✉)
Head of Tissue Services, NHSBT Tissue Services, Liverpool, L24 8RB, UK
e-mail: John.Kearney@nbs.nhs.uk

G. Galea (ed.), *Essentials of Tissue Banking*, DOI 10.1007/978-90-481-9142-0_7,

Autologous Versus Allogeneic Grafts

In almost all cases the use of autologous tissue i.e. a graft taken from elsewhere in the body to replace diseased, damaged or missing tissue is the gold standard. Because autografts are "self" they will not elicit an immunological rejection response. If used immediately after harvest they will be fully viable and hence cells within the graft will contribute to the repair process. Indeed for vascularised grafts the vessels within the graft will often enosculate with vessels in the surrounding wound bed thus ensuring rapid re-perfusion of the grafts.

However there are limitations to the use of autografts. In particular, it is important not to create a non-healing defect at the donor site(s). So for example with bone, this means taking only small samples, below the critical size, that will allow natural regeneration, whereas for skin it means taking only partial thickness grafts or very small full thickness punch biopsy grafts. Even then, the patient may suffer significant morbidity at the donor sites including pain, infection, etc. In addition, for certain tissue there are no suitable autografts available e.g. the body only has a single aortic heart valve, therefore a defective aortic valve cannot be replaced by an alternative autologous aortic valve.

For these reasons the practice of tissue allografting developed as an adjunct to autografting, or as the only option where a suitable autograft was not available. Although in principle an allograft transplant could be carried out in one operation, by the retrieval of viable tissue from a living donor (or deceased donor on life support) and transfer directly into the recipient i.e. analogous to an autograft; in practice for various reasons, this is rarely undertaken. The use of a viable rather than a non-viable allograft, may in fact be detrimental and hence this level of organisational urgency unwarranted.

Immunological Reponses to Tissue Allografts

Any tissue or organ that contains living cells is likely to elicit an "acute allograft rejection response" via one pathway or another. The ultimate outcome is that the allogeneic cells will be killed by the recipient's immune response. For organs or tissues that are highly cellularised, and vascularised, the outcome is rapid killing of the vessel endothelial cells leading to thrombosis, vessel occlusion, and death and necrosis of the other cell populations i.e. classical rejection of an organ or skin graft, with easily observable macroscopic changes. In contrast, a piece of cortical bone containing scanty osteocytes may well elicit an acute allograft rejection response owing to stimulation by allogeneic passenger leukocytes, but the consequence is unlikely to be noticeable macroscopically. Indeed, the osteocytes may not be immediately killed but rather opportunistically killed when the bone resorption/replacement process brings the cells in close proximity to the vascular system, which delivers the recipients immunological effectors (leukocytes and antibodies) to the allogeneic cell. Therefore for this tissue, "immunological rejection" is unlikely to be apparent nor is it likely to interfere with the tissue regeneration process whereas for the previous examples it would totally destroy the tissue and any prospect of regeneration.

Many other tissues are between these two extremes. Cancellous bone, for example, contains very few "bone cells" but huge quantities of bone marrow and fat within the trabeculae. These cells would elicit a huge rejection response that would at least delay the regeneration response and may cause some collateral damage or longer lasting inhibition of graft recolonisation by tissue regeneration cells [2]. It would therefore be good practice to remove at least the trabecular marrow cells prior to grafting. Removing the fat would also accelerate graft incorporation [3].

The only way to prevent the rejection of allogeneic tissue cells is with HLA matching and immunosuppression, as happens with organ grafts. This has been evaluated for skin grafts for use in burns [4, 5]. However as most tissue grafting is carried out as a life enhancing rather than life saving operation, and there are significant morbidity and mortality consequences of using intensive immunosuppression, this approach has not been generally adopted.

Therefore, where the viable donor cells will eventually be killed by the immune response, the only circumstances where using a viable tissue allograft will have benefit is if the graft performs a vital function prior to rejection. One example is the use of viable skin allografts. These will gain a blood supply by enosculation with vessels in the wound bed and thus provide true biological closure of the wound and fully functional skin. This buys the surgeon some time to find additional permanent grafts (e.g. by re-cropping donor sites at intervals or growing cultured autografts). In addition very severely burned patients are to some extent immunosuppressed as a result of the injury so rejection may be significantly delayed, buying even more time. The use of viable skin allografts can therefore be life-saving for severely burned patients.

For most other tissues, there is no benefit to using a viable graft. Tissue banking science has therefore developed methods for banking of tissues in a non-viable state which then allows application of other beneficial processes such as sterilisation to avoid the risk of transmission of disease from donor to recipient or as a result of contamination during processing.

There are however certain "privileged" tissues that do not elicit an "acute allograft rejection response" even though they are fully viable. These are tissues that are not normally vascularised and hence do not become vascularised after grafting. The vascular system is the conduit along which donor cells travel to the lymphoid tissue to elicit an immune response (the afferent arm) and along which recipient effector cells (and antibodies) travel back to the tissue to destroy the donor cells. In the absence of a vascular network, neither can occur. Two examples of avascular tissues are articular cartilage and cornea. Therefore, in most cases, these tissues can be used as viable allografts without being rejected, and indeed do not function adequately or regenerate if non-viable grafts are used. Therefore development of techniques for banking of viable tissues has also been an important objective for tissue banks. Cryopreservation (or vitrification) is the commonest technique used for long term storage. This is the subject of a separate chapter in this book and will not be covered here. However other short term banking methods for viable tissues will be briefly reviewed.

Principles of Tissue Preservation

Preservation of Non Viable Tissue Grafts

The challenge of tissue preservation is one that has faced mankind for millennia. Animal and plant tissues are a major component of the human diet. Both are subject to progressive degradation once harvested therefore methods to prevent degradation of food have been essential to survival. These methods have included drying of the tissue, use of low temperatures and the use of chemical preservatives such as salt. Although these techniques developed by trial and error, the underlying principles are now better understood. The more recent discipline of "Tissue Banking" has borrowed heavily from this knowledge.

Causes of Degradation

One of the major potential causes of tissue degradation is the growth and activity of microorganisms. Whereas in life there are mechanisms to exclude and where necessary, kill microorganisms to prevent their access to tissues; once tissues are removed from the body these mechanisms rapidly disappear. However, as microorganisms that are transferred to recipients can cause serious diseases, methods have been developed to either avoid contamination or to inactivate microorganisms. These methods will be discussed in a separate section later in this chapter. However, it is well known that microorganisms can only grow in the presence of water.

The action of enzymes can also cause damage to both the cells and the matrix of stored tissue grafts. The elaboration of degradative enzymes in life is an important part of tissue turnover, the removal of dead/damaged tissue and in the process of apoptosis. Removal of tissue from the body results in cellular necrosis and the release of degradative enzymes. The presence of water is essential for enzyme activity.

Some of the most powerful degradative enzymes are the hydrolytic enzymes, found for example in lysozomes. Hydrolysis can also be caused by extremes of pH together with raised temperature. As the name suggests, water molecules are essential to the process.

Lipid peroxidation is another potent degradation process. This involves reactions of lipids with reactive oxygen species or "free radicals". Once started, the process continues as a chain reaction. It leads to rancidity of fats in food, but can result in cytotoxic by-products in tissue grafts [6, 7]. In the presence of water, radical formation generates the most reactive species, the hydroxyl radical (OH°).

It is clear from the foregoing that all of the degradative reactions that can adversely affect the preservation of tissue grafts, are dependant on the presence of water. This water must be in a free state and able to react i.e. must not be immobilised.

The relationship between degradation rate for each of these processes and water activity is shown in Fig. 7.1.

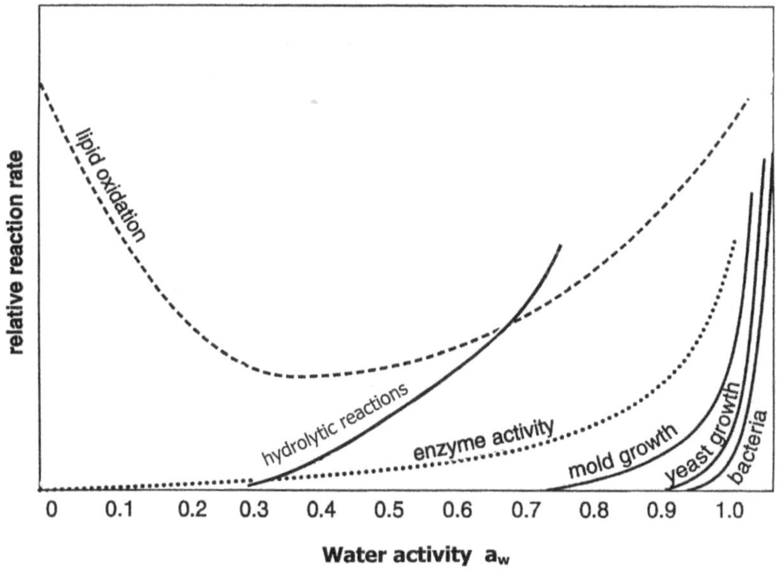

Fig. 7.1 The relationship between water activity and detrimental microbial and chemical reactions. Adapted from *Cell Tissue Bank* [11]

Water activity ranges from aw = 1 which is 100% free water, to aw = 0. As the water activity decreases, first bacteria then yeasts have reduced growth rates decreasing to zero at aw = 0.9. Moulds are better able to tolerate reduced water activity, however at aw = 0.7 they also cease growth activity.

Enzyme activity decreases rapidly initially but not totally disappearing until aw = 0. Hydrolytic reactions are more tolerant of lower aw initially but approach zero at aw = 0.3. Likewise lipid peroxidation reaches a minimum at aw = 0.3 but increases either side of this. Therefore, for tissue preservation, an ideal water activity range would be circa aw 0.2–0.4. This would minimise the major degradation reactions.

Reduction of Water Activity

Water activity can be reduced in a number of different ways. The water can be physically removed by drying the tissue matrix, or by immobilising the water so that it is no longer "active" and hence does not contribute to water activity.

Freeze-Drying (Lyophilisation)

If tissues are dried at ambient temperature, the progressive loss of water results in an increase in the salt concentration in the remaining water in contact with the tissue matrix. This can lead to conformational changes in molecules and to denaturation. In contrast, if the tissue is first deep frozen, and the water removed by sublimation by applying a vacuum, this will prevent molecular unfolding. Once dried, the tissue can be returned to room temperature without risk of molecular unfolding or degradation reactions, as both require the presence of water.

During the freeze drying process the tissue remains frozen within a refrigerated chamber. A vacuum is then applied to the chamber across a condenser, the condenser being at a lower temperature than the tissue chamber. In this way water molecules sublime and are transferred from the chamber to the condenser [8]. As the process continues water is lost from the tissue much more slowly, therefore the temperature of the tissue chamber is gradually increased. Using this process a aw of 0.2–0.4 can easily be achieved.

Deep Freezing

One method to "immobilise" the water within a tissue is deep freezing. The water enters the solid phase as ice crystals and hence is no longer available as free water to take part in degradation reactions. The lower the temperature, the lower the water activity. In addition low temperature itself prevents microbial growth and reduces reaction rates of eg enzymatic reactions. In practice, storage temperatures below –40°C or ideally –80°C are suitable for the long term storage of tissue matrices [7, 9].

High Concentration Solutes

The major solutes used for preservation of food stuffs have included salt and sugar (in preserving eg fish and fruit respectively). In contrast, glycerol has been the major solute used for tissue graft preservation, presumably owing to its longstanding safe use as a cryoprotective agent for tissues and cells. The principal for the immobilisation of water by solutes, is that each solute molecule will sequester water molecules within a hydration shell. The more solute molecules, the more water that becomes immobilised in the hydration shell hence the less free water.

The Euroskin Bank developed a glycerol, solute-preservation method for preserving non-viable skin whereby the skin was incubated successively in 50, 70 and then 85% glycerol [10]. Further, more detailed, characterisation of the glycerol and water fluxes in this system have yielded a more efficient and fully validated protocol [11–13].

It should be noted that it is important to fully remove these high concentrations of glycerol from the grafts prior to clinical use, to avoid high systemic concentrations of glycerol in the patient. At high doses glycerol does have toxic effects on muscles causing myonecrosis which in turn can result in renal failure and even death [14, 15]. Toxicity studies have shown that the LD 50 for glycerol is 0.00442 mL/g [16].

Fully validated protocols for the removal of glycerol from skin have been developed [11]. This is achieved by repeatedly washing the skin in physiological saline for at least 30–60 min.

Cell Removal

For tissues that are stored as non-viable grafts, the dead allogeneic cells serve no useful purpose, and in fact are reservoirs of degredative enzymes and immunogenic molecules. This has led some groups to develop methods to remove all cells and cell components from the tissue prior to preservation and subsequent clinical use. It is important that the decellularisation process does not adversely affect the biomechanical and biological properties of the tissue matrix. At least some decellularised tissues have been shown to incorporate and re-model more rapidly and to avoid foreign body responses by the recipient [17].

Preservation of Viable Tissue Grafts

The preservation of viable tissue grafts is even more challenging than for non-viable tissues. Once removed from the body and hence deprived of perfusion by the vascular system, the metabolising cells within the tissue become dependant on diffusion for the delivery of nutrients and oxygen and the removal of metabolites. For tissues bathed within a nutrient medium at normothermic temperature (circa 37°C) diffusion from the surface of the tissue into its depths is a very inefficient process.

If cells within the central mass of the tissue are deprived of nutrients and oxygen for prolonged periods they will die and a central necrosis of the tissue will commence. The release of degradative enzymes will then cause the various degradative processes described previously for non-viable tissue, to occur. For tissues where cells are only located on the outer surfaces of the graft eg eye cornea, comprising of an outer epithelial layer and an inner endothelial monolayer, diffusion is adequate for maintaining cell viability even when maintained at normothermic temperatures; but for most other tissues reliance on diffusion is inadequate for periods longer than a few hours.

By reducing the storage temperature from normothermic (circa 37°C) to hypothermic (circa 4°C), the metabolic rate of the cells will be reduced and hence the demands for nutrients and oxygen. This has been shown to reduce the rate at which tissues lose viability [18]. However, at this lower temperature certain essential biochemical pathways may be inhibited. One example is the Na^+/K^+/ATP pump which maintains higher levels of K^+ and lower levels of Na^+ within cells than in the extracellular medium. To prevent the influx of sodium under hypothermic temperatures, some authors have recommended using a bathing salt solution formulated to reflect intracellular concentrations of Na^+ and K^+ [19].

Another reason why the viability of tissues declines rapidly when isolated from the body and maintained at lower temperatures, may be the effects of free radical damage. In life, the body has many systems to scavenge or inactivate free radicals which would otherwise attack cell membranes and macromolecules. When isolated from the body and maintained at reduced temperature, it is unlikely that the tissue will continue to produce sufficient antioxidants and scavenging enzymes to protect the cells and cell components.

All of these mechanisms combine to render viable non-frozen storage of most tissue grafts as being a very time limited method of storage.

Principles of Disinfection and Sterilisation

The major strategy for avoiding transmission of viral and other infectious diseases from the donor to the recipient is the rigorous screening of the donor's medical and behavioural history, and testing of blood samples for the presence of viral nucleic acid (NAT) or an immunological response to the virus (antibody detection). However, there are opportunistic bacterial and fungal pathogens that may contaminate the tissue graft after death, during procurement or during processing within the tissue bank. In addition, certain tissues (e.g. skin and the gut) are associated with a resident microflora in life. Some of these microorganisms may become pathogenic in a surgical wound situation.

Although many tissue banks attempt to avoid contamination of grafts by applying strict asepsis during procurement and subsequent processing, the "sterile" nature of the graft can only be confirmed by final sampling of the tissues for microorganisms. Because destructive sampling of the whole graft defeats the purpose, most sampling regimens are only likely to detect gross contamination. It is reasonable to assume that an immunocompetent patient can eradicate small numbers of bacteria from a well vascularised tissue graft. The problem arises where the graft tissue is not well vascularised and areas of ischemia or necrosis occur. This produces anaerobic conditions ideal for the growth of eg Clostridium species, which can be highly pathogenic. Just a few Clostridial spores could lead to a life threatening disease. This has led many tissue banks to routinely apply disinfection or sterilisation techniques to inactivate bacteria and fungi. Clearly, it is important that these techniques do not significantly damage the tissue graft or adversely affect its function or efficacy.

Definitions

"Disinfection" is a nebulous concept referring to the removal from an environment of microbes that might cause disease. In contrast "sterilisation" is now framed as a statistical concept involving the destruction of all microorganisms with a defined level of probability.

For any given population of microorganisms, the application of a sterilising chemical or physical agent will kill the microbial cells as an exponential function, to a first approximation (Fig. 7.2).

When the number of microorganisms is transformed to log 10_{10} this approximates a straight line (Fig. 7.3).

Therefore the dosage or exposure time required to reduce a given population, tenfold (i.e. 1 $\log_{10}$) is a constant equal to the gradient of the line. This is defined

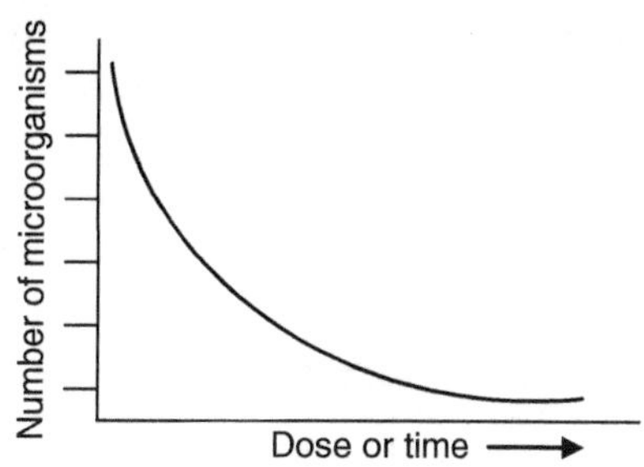

Fig. 7.2 Exponential decrease of microorganisms with increasing exposure to sterilant

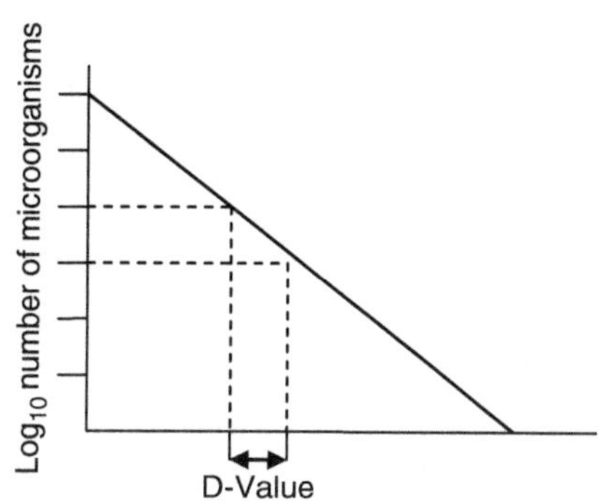

Fig. 7.3 The D-value concept

as the D-value (decimal reduction value). Using the D-value concept it is possible to predict the probability of a surviving microorganism after application of a defined amount (or exposure time) to a sterilising procedure; providing the starting level of contamination (the bioburden) is known. In the pharmaceutical industry the accepted level of sterilility assurance (SAL) is to have no more than 1 viable microorganism per million units of product, i.e. a SAL of 10^{-6}. Therefore if a product has a starting contamination level of 10^3 microorganisms per unit, and a SAL of 10^{-6} is required, then the sufficient dose of sterilant required is $9 \times$ D-value (Fig. 7.4).

If there is more than one species of microorganism present, all at different contamination levels, and with different susceptibilities to the sterilant, then the analysis becomes more complex. Two approaches have been used. One is to assume a "worst case" scenario i.e. take values for the most resistant microorganism and choose a worst case contamination level. If a SAL of 10^{-6} is still achieved then the tissue can be considered "sterile" [13]. An alternative method using actual bioburden levels, and average susceptibility levels of each microorganism has been developed [20]. Once a sterilisation method has been fully validated, it should be possible to use parametric release i.e. if it can be shown that the sterilisation load was subjected to the defined level of the sterilisation process, then sterility can be assumed. In contrast, for disinfection, it is necessary to always test a sample of the tissue for the presence of residual pathogens.

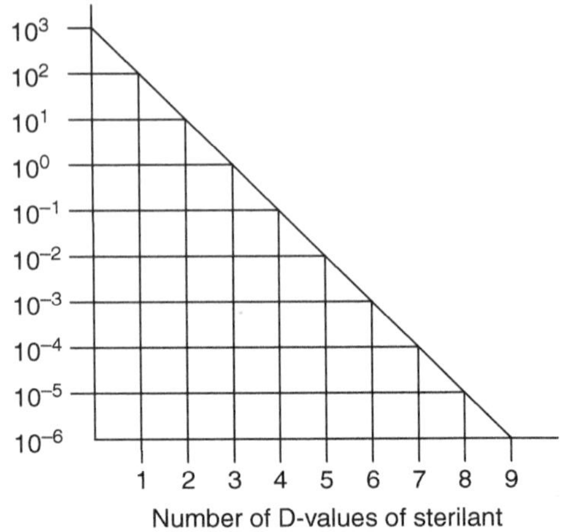

Fig. 7.4 Number of D-values of sterilant required to reduce bioburden from 10^3 to 10^{-6}

Disinfection Methods

Viable Tissues

For viable tissue grafts the most common disinfection method is the use of antibiotics/antimycotics. These agents are used systemically for therapy. They may either kill the microorganism (bacteriocidal) or prevent its growth (bacteriostatic) so that the immune system can locate and kill the microbe. When used in vitro to disinfect tissue grafts such as heart valves, any bacteriostatic effect is not particularly effective as there is no immune system to clear the microbes and once implanted the concentration of the antibiotic will rapidly be diluted thus allowing microbial growth. Therefore it is the bacteriocidal effect which is most beneficial in vitro.

To ensure effectiveness against a range of microorganisms, a cocktail of antibiotics/antimycotics is often used. Care should be taken to fully validate the use of combinations of antibiotics as both synergistic and antagonistic effects may be seen, so that the published Minimum Bacteriocidal Concentration (MBC) for different microbes may no longer be relevant. In addition, the therapeutic use of antibiotics takes place at normothermia (i.e. 37°C for systemic delivery and slightly lower for topical use). In contrast, many protocols advocate the use of antibiotic cocktails at ambient (circa 20°C) or refrigerator (circa 4°C) temperatures. Care should be taken to use antibiotics with a mode of action relevant at these temperatures. For example Penicillin acts by preventing the normal synthesis of cell walls during bacterial growth and therefore requires bacteria to be actively proliferating. At 4°C few, if any, pathogenic bacteria are able to grow, hence Penicillin is unlikely to be effective at this temperature.

Caution should be taken when using antimycotics. Fungi, like humans, belong to the Eukaryotes whereas bacteria are Prokaryotes. There are many more drug targets against Prokaryotes not shared with humans, than there are for fungi. Therefore whereas antibiotics can be profoundly toxic to bacteria but hardly at all towards human cells, even at high concentrations, this is often not the case for antifungals [21]. Therefore any novel cocktail should be fully validated for cytotoxicity against critical tissue cells prior to its introduction.

Non-viable Tissues

Generally, if a tissue is to be used as a non-viable tissue implant, application of a sterilisation technique would be preferable. However, there are sometimes concerns surrounding the effects that harsh sterilisation techniques might have on the tissue structure and efficacy. In these cases a less severe disinfection is common. Ethanol is rapidly lethal to non-sporulating bacteria and destroys mycobacteria but is ineffective at all concentrations against bacterial spores.

In addition to ethanol, other alcohols, organic solvents and detergents have been used to disinfect tissues. Specific combinations have been developed to produce proprietary "sterilisation" methods for banked tissues, where a SAL of 10^{-6} is claimed.

Other antiseptics and disinfectants used in the medical setting eg hydrogen peroxide and other peroxygens, chlorine compounds and other halogens and even mercurial compounds such as Cialit [22] (used to preserve middle-ear ossicles) have been used to disinfect tissue grafts.

Mild heat treatment has been used to inactivate vegetative bacteria but care must be taken to avoid destroying advantageous heat labile components (such as growth factors), and denaturation of structural proteins such as collagen. Because the collagen in bone tissue is protected by a mineral coating, this tissue can withstand higher temperatures before denaturation commences. A temperature of 80°C was generated in a device developed for the disinfection of femoral heads [23], but many bacteria are fairly resistant to these temperatures whereas beneficial biochemicals such as the BMP's are heat labile.

Sterilisation Methods

Physical Methods

With respect to physical methods of sterilisation, the temperature required for dry heat sterilisation (over 160°C) or wet heat sterilisation (in excess of 121°C) are detrimental to the structure of tissue. Even bone tissue with its protective mineralised coating was shown to function poorly in vivo after boiling [24]. The only physical method currently widely used for tissue sterilisation is the use of irradiation (either gamma – or electron beam). This is the subject of another chapter in this book and will not be considered further here.

Chemical Methods

For many years ethylene oxide was the preferred chemical sterilant. When used as a gas it was shown to effectively penetrate tissue matrices including dense cortical bone [25]. More recently however, there has been concern about the genotoxicity of residual gas left in the matrix, for which there is thought to be no safe residual level. Glutaraldehyde has been advocated as a sterilising chemical, and has been used in the manufacture of porcine heart valves [26]. However, in addition to killing microorganisms it also significantly cross-links the collagen within the tissue. This renders porcine tissue much less immunogenic but also causes stiffening of the tissue including the valve leaflets which is disadvantageous and induces rapid calcification in vivo. Peracetic acid is now frequently used in hospitals for sterilising endoscopes in preference to glutaraldehyde. It has also been used for sterilising bone grafts [27]. Its use for other tissues must be undertaken with care as tendons treated with relatively high concentrations of peracetic acid were shown to have inferior performance in vivo [28]. In contrast, heart valves sterilised using low concentrations of peracetic acid were shown to function satisfactorily in a sheep model (E. Ingham personal communication). Other chemical sterilising agents used in hospitals and medical manufacturing, such as vapour phase hydrogen peroxide, ozone, etc, have not been extensively evaluated for tissue grafts.

References

1. Wasiak J, Clark C, Villanueva E (2008) Autologous cartilage implantation for full thickness articular cartilage defects of the knee (review). The Cochrane Library: Issue 4
2. Burwell RG (1963) Studies in the transplantation of bone V. The capacity of fresh and treated homografts of bone to evoke transplantation immunity. J Bone Joint Surg 45B:366–401
3. Aspenberg P, Thoren K (1990) Lipid extraction enhances bank bone incorporation: an experiment in rabbits. Acta Orthop Scand 61(6):546–548
4. Achauer BM, Black KS, Waxman KS et al (1986) Long-term skin allograft survival after short-term cyclosporin treatment in a patient with massive burns. Lancet 1(8471):14–15
5. Dreno B, Meignier M, Bignon JD et al (1987) Immunological mechanisms of cyclosporin in skin allograft. Lancet 2 (8570):1270–1271
6. Moreau MF, Gallois Y, Basle MF et al (2000) Gamma irradiation of human bone allografts alters medullary lipids and releases toxic compounds for osteoblast-like cells. Biomaterials 2(4)1:369–376
7. Laitinan M, Kivikan R and Hirn M (2006) Lipid oxidation may reduce the quality of a fresh-frozen bone allograft. Is the approved storage temperature too high? Acta Orthop 77(3): 418–421
8. Kearney JN (1996) Banking of skin grafts and biological dressings. In: Settle JAD (ed) Principles and practice of burns management. Churchill Livingstone, pp 329–351
9. Guidelines for the Blood Transfusion Services in the United Kingdom 7th edn 2005. The Stationery Office, London
10. Kreis RW, Vloemans AFPM, Hoekstra MI et al (1989) The use of non viable glycerol-preserved cadaver skin combined with widely expanded autografts in the treatment of extensive third-degree burns. J Trauma 29:51–54
11. Huang Q, Pegg DE, Kearney JN (2004) An improved glycerol banking method used in the preservation of non viable skin allografts. Cell Tissue Bank 5:3–21

12. Ross A, Kearney JN (2004) The measurement of water activity in allogeneic skin grafts preserved using high concentration glycerol or propylene glycol. Cell Tissue Bank 5:37–44
13. Kearney JN (2005) Guidelines on processing and clinical use of skin allografts. Clin Dermatol 23:357–364
14. Backenroth R (1998) Glycerol induced acute renal failure attenuates subsequent $HgCl_2$ – associated nephrotoxicity: correlation of renal function and morphology. Ren Fail 20:15–26
15. Zurovsky Y (1993) Models of glycerol-induced acute renal failure in rats. J Basic Clin Physiol Pharmacol 4:213–228
16. Uche EM, Arowolo RO, Akinyemi JO (1987) Toxic effects of glycerol in swiss albino rats. Res Commun Chem Pathol Pharmacol 56:125–128
17. Mirsadraee S, Wilcox HE, Watterson KG et al (2007) Biocompatibility of acellular human pericardium. J Surg Res 143:407–414
18. Rosenquist MD, Cram AF, Kealey GP (1988) Skin preservation at 4°C: a species comparison. Cryobiology 25:31–37
19. Taylor MJ, Hunt CJ (1985) A new preservation solution for storage of corneas at low temperatures. Curr Eye Res 4(9):963–973
20. Yusof N (1999) Quality system for the radiation sterilisation of tissue allografts. Adv Tissue Bank 3:257–281
21. Agurregoicoa V, Kearney JN, Davies GA et al (1989) Effects of antifungals on the viability of heart valve cusp derived fibroblasts. Cardiovasc Res 23(12):1058–1061
22. Betow C (1982) 20 years experience with homografts in ear surgery. J Laryngol Otol Suppl 5:1–28
23. Von Garrel T, Knaepler H (1999) Surgical femoral head allograft processing system using moderate heat. Adv Tissue Bank 3:283–354
24. Burwell RG (1966) Studies in the transplantation of bone VIII. Treated composite homograft-autografts of cancellous bone: an analysis of inductive mechanisms in bone transplantation. J Bone J Surg 48B:532–566
25. Kearney JN, Bojar R, Holland KT (1993) Ethylene oxide sterilisation of allogeneic bone implants. Clin Mat 12:29–33
26. Butterfield M, Fisher J, Kearney JN et al (1991) Hydrodynamic function of second generation porcine bioprosthetic heart valves. J Card Surg 6(4):490–498
27. Pruss A, Kao M, Kiesewetter H et al (1999) Virus safety of avital bone tissue transplants: evaluation of sterilization steps of spongiosa cuboids using a peracetic acid-methanol mixture. Biologicals 27:196–201
28. Scheffler SU, Gonnermann J, Kamp J et al (2008) Remodelling of ACL allografts is inhibited by peracetic acid sterilization. Clin Orthop Relat Res 466(8):1810–1818

Chapter 8
Cryopreservation

David Pegg

Introduction

"Cryopreservation" is sometimes said to provide "viable" tissue – the terms have even been used as synonyms [1]. But the fact is that a cryopreservation method that is effective for one tissue may not be effective for another and even if reasonably effective it may not necessarily produce fully functional, living tissue – some cells may have been destroyed. One sometimes encounters the oxymoron "preservation injury" but surely, "preservation" and "injury" are mutually exclusive. And what does the term "viability" mean in a scientific context? The situation is complicated but at least some of the confusion may be resolved if we can define some of these terms.

Literally, "cryopreservation" means preservation by reduction of temperature. However we all know that cooling living cells and tissues much below zero on the Celsius scale causes freezing, and we also know that freezing severely damages cells and is even used to deliberately destroy tissue in cryosurgical procedures. Profound cooling will be preservative only if the damaging effects of freezing can be avoided or at least greatly minimised. In 1949, Polge, Smith and Parkes [2] discovered that spermatozoae would survive prolonged freezing at –80°C if they were suspended in a saline solution containing glycerol; the sperm were preserved effectively because a way had been discovered to prevent freezing injury: the metabolic slow-down produced by cooling could then be used to preserve living cells. The term "cryopresevation" was adopted to describe methods that avoided ice-damage leaving just the preservative effect of profound reduction in temperature. There are essentially two approaches to achieving this situation. The first is to minimise the amount of ice that is formed and restrict ice crystals to parts of the sample where they are harmless or nearly so. This is the approach that Polge, Smith and Parkes hit upon and now, 60 years later, it is commonly referred to as "conventional cryopreservation". The other approach is to cool to very low temperatures in such a way that

D. Pegg (✉)
Department of Biology (Area 14), University of York,
York YO10 5YW, UK
e-mail: dep1@york.ac.uk

G. Galea (ed.), *Essentials of Tissue Banking*, DOI 10.1007/978-90-481-9142-0_8,

the crystallization of ice is prevented altogether; this is called "vitrification". The essence of both types of "cryopreservation" is that the cells or tissues are stabilised at a very low temperature in such a way that any significant ice damage – whether direct or indirect- is eliminated and the material can be restored to its physiological state with a high degree of functional recovery. As will be discussed later, cells and tissues differ in their requirements for survival. Hence, a cryopreservation method that works well for one tissue (for example small elastic arteries) [3] may not work at all for another (for example articular cartilage) [4]. It is important to realise that cryopreservation methods can be specific to particular cells and tissues and the use of such a proven cryopreservation method with different cells or tissues does not guarantee effectiveness.

Consider next the term "viability". There is much imprecise and confusing use of this word in the tissue banking world: in the dictionary, as in obstetrics, it means "capable of life" whereas it is usually used – in a "viability test" – to indicate whether the cells in question actually are alive. Many so-called "viability tests" have been invented, such as the Trypan Blue permeability test or propidium iodide staining. Some are misleadingly called "live/dead" stains. But these are not "viability" tests; all they do is test whether the cell membrane is leaky to a particular dye so these assays are more properly called membrane integrity tests. A particularly important error is to imply, usually without an explicit statement, that satisfactory performance in one test denotes the presence of all the characteristics of life. That just is not true [5]. Assays of vitality should be designated by what they actually measure – response to a pharmacological agonist [3] or ability to synthesise a characteristic molecule [4].

The crucial question in tissue banking is, "What properties must the preserved tissue possess if it is going to work for the patient and surgeon?" That may mean that it has to contain physically intact, metabolising cells that are capable of cell division (for example haemopoietic stem cells) or it may not. It will often mean that other properties, that have nothing to do with whether the cells are alive, are essential. Therefore, tissues for tissue banking purposes can be classified as those that do require the presence of cells that exhibit the formal characteristics of life (growth, replication, locomotion and irritability) in order to function as a graft...and those that don't. The latter may not normally be considered in a chapter on cryopreservation but they will be considered here, very briefly, because they do rely on reduction in temperature for preservation.

Tissues that Do Not Require Living Cells for Effectiveness

The most commonly banked tissue is bone. The crucial properties that are required are mechanical – adequate strength and lack of brittleness. In addition, bone may need to retain some biochemical characteristics, for example the bone morphogenic protein which is responsible for promoting the regeneration of new bone in the recipient So, both mechanical and biochemical properties must be retained but living cells are not required. As a result, bone is commonly preserved simply by freezing and storage at –80°C, or by freeze-drying. Water is still within tissue that is frozen at –80°C but it is crystallized as ice whereas in freeze-drying it is actually removed. Tendons and fascia do not require living cells either but the collagen matrix must retain adequate properties. These tissues are also commonly stored at –80°C.

Tissues that May Not Require Living Cells

There are other tissues for which there is a debate concerning the need for living cells. Skin is such a tissue. Some surgeons argue that skin does not need live donor cells in order to be effective for the treatment of burns or ulcers; it is however important that the collagen matrix is intact and enables the host's cells to grow and repair the damage with minimal scarring and contraction. Preservation is achieved by immersion in a very high concentration of glycerol at +4°C. The theoretical basis of this method is unclear but it can give satisfactory clinical results. One practical problem is that glycerol diffuses very slowly and removing it when the tissue is required may be a problem. We have suggested that propylene glycol might be a satisfactory alternative, bringing the advantage that this solute diffuses much more rapidly than glycerol [6]. The other approach is to use a cryopreservation method that does enable at least some of the donor cells to survive. However, it does not appear to have been established that cryopreserved skin is clinically superior to glycerolised skin. From a theoretical viewpoint there is also the potential disadvantage that live donor cells may stimulate cell-mediated rejection by the recipient.

Cardiac valves are another tissue graft about which there is debate. There are strong advocates of the so-called "viable valve graft" [7] but scientific evidence for the survival of significant numbers of donor cells in cryopreserved cardiac valves is weak: Nor has it been convincingly shown that such cells have any beneficial effect for the graft recipient. In fact, whenever valve grafts are removed subsequently they are always grossly hypocellular although they may have been functioning perfectly well. We have reported some experiments in which sheep heart valves were preserved using the same glycerol method that is used for skin, so there were certainly no living cells in these grafts when they were grafted into sheep [8, 9]. Two other experimental groups in this study were untreated controls or valves that had been cryopreserved by a standard method that we had shown to allow living cells to survive. After 6 months the groups were indistinguishable from each other. We found no strong evidence that cardiac valve grafts need to contain living cells though an intact connective tissue structure certainly is essential.

Tissues that Do Require Living Cells

In contrast to the above, there are many examples of cells that must be preserved in a fully functioning state and similarly there are tissues that require the constituent cells to be alive and to function. Some of these can be preserved by conventional cryopreservation methods and some cannot. A good example of cells that most definitely must be fully functional is provided by haemopoietic stem cells whether from bone marrow, peripheral blood or cord blood. Only 6 years after the discovery of the cryoprotective effect of glycerol, Barnes and Loutit [10] showed that a similar technique could be used to preserve mouse haemopoietic stem cells, and that the cryopreserved cells could then be used to rescue animals that had been exposed to a lethal dose of ionising radiation. Haemopoietic stem cells provide a good example for discussion of the principles of cryopreservation

Cryopreservation

There are certain basic principles to be followed when designing a cryopreservation method for any type of cell. These will now be discussed in a step-by-step sequence.

Step 1: is to select a cryoprotective agent (CPA). Glycerol was the first widely used cryoprotectant but in 1959 Lovelock and Bishop [11] introduced dimethyl sulphoxide (DMSO) as an alternative that had the advantage diffusing more rapidly than glycerol, and hence with less osmotic stress upon the cells. Lovelock listed the requirements of cryoprotection as: high solubility; an ability to produce a profound depression of freezing point; ability to penetrate cells; and low toxicity [12]. Such compounds were cryoprotective because they reduced the amount of ice formed at any given subzero temperature and therefore moderated the rise in salt concentration that accompanied the sequestration of water in the growing ice crystals. The last two requirements (ability to penetrate cells and low toxicity) interact because solutes that permeate rapidly require briefer exposure times and produce less osmotic disturbance; consequently a greater intrinsic or chemical toxicity may be tolerated. Thus, glycerol is intrinsically less toxic than DMSO but permeates much more slowly, as a result of which exposure times have to be longer and osmotic stress is less easily controlled. In general, DMSO is the preferred cryoprotectant and Ashwood-Smith [13] reported that irradiated mice, infused with similar doses of bone marrow cells that had been cryopreserved either with glycerol or with DMSO recovered more rapidly when DMSO was used as the cryoprotectant. Unless there are good reasons for doing otherwise the first choice of cryoprotectant will be normally be DMSO. However, this choice should be reviewed after completion of the whole cryopreservation protocol.

All penetrating cryoprotective agents (CPAs) act by reducing the salt concentration that otherwise increases during the freezing process, and because the CPA is present inside as well as outside the cells, it controls both intracellular and extracellular salinity. Non-penetrating solutes can also have cryoprotective effects, but in most applications they are less effective than penetrating cryoprotective agents. They do control the build-up of extracellular salinity but have no effect on intracellular salt concentration; the details of their mode of action are still unclear. Although they are generally less effective per se, non-penetrating solutes such as polyvinylpyrrolidone (PVP) and hydroxyethyl starch (HES) have sometimes been used in combination with penetrating agents such as DMSO.

Step 2: is to design the method of adding and removing the CPA such that the cells are not subjected to serious osmotic stress or chemical toxicity: this requires measurement of three properties. The first is the cell's volumetric response to changes in external osmolality using impermeable solutes; the object is to identify the upper and lower limits of swelling or shrinkage that the cells will tolerate, as indicated by an appropriate functional test. These data are obtained by measuring cell volume and estimating cell damage in solutions of a non-toxic, non-penetrating solute of differing but known osmolality. Sucrose or mannitol are commonly used and volume may be measured by use of a Coulter counter [14] or by an optical method [15]. The relationship between cell volume and osmolality is a straight line when volume is regressed on the reciprocal of osmolality and the intercept on the y-axis defines volume at infinite osmolality, the so-called non-osmotic volume. This is known as a Boyle van't Hoff plot, and such data for human cord blood stem cells have been reported by Hunt, Armitage and Pegg [16]. The proportional non-osmotic volume was found to be 0.27 for these cells. The other two parameters that are required for the design of the addition/removal protocol are the water permeability (or hydraulic conductivity, L_p) and the solute permeability of the CPA (P_s). These data are calculated from the time course of measured cell volume following exposure to a known concentration of the chosen CPA at the chosen temperature. The theory behind these measurements is described elsewhere [17] and the necessary software can be run on a regular PC. Plugging these parameters into the relevant equations enables us to calculate the maximum transient decrease in volume that will follow exposure to any cryoprotectant solution of known composition, the maximum increase in volume that will follow immersion of the CPA-loaded cell in a solution containing a lower concentration of CPA and the time course of these processes at selected temperatures, say +4 and +22°C. If the tolerated limits of volume excursion are known then the acceptable step changes in concentration can be calculated. In the case of cord blood stem cells the limits were found to be 45–140% of isotonic volume. It was possible to add DMSO to a final concentration of 1.4 molar in a single step and to remove it in three steps over an 80 min period without transgressing the tolerated volume limits.

Step 3: uses the chosen steps in concentration and the selected time and temperature of exposure to determine the greatest tolerated concentration of CPA under these conditions of exposure. Thus, experiments are done in which the maximum concentration of CPA is progressively raised; the cells are allowed to remain at that concentration for, say 30 min; the CPA is then removed by the decreasing sequence of concentration steps; and a suitably sensitive and appropriate assay is applied. In this way the highest acceptable concentration of CPA can be determined. Note the words *sensitive* and *appropriate*: membrane integrity tests are hopelessly insensitive for this purpose. An appropriate test is one that measures a function that the cell is required to perform in the intended application – in the case of human cord blood haemopoietic cells an assay of colony formation in culture would be suitable. For DMSO the final concentration that has most commonly been used is 10%v/v.

Step 4: is to cool the cell suspension to the intended storage temperature, in the presence of the chosen CPA added as defined in steps 1–3, store, then thaw, remove the CPA and assay for functional recovery. This step requires selection of a cooling trajectory (a temperature/time profile), a storage temperature and a thawing method. This raises theoretical issues that are discussed elsewhere [17]: current theory is not sufficient to predict with certainty just what the optimal cooling and warming rates might be. Determination by direct experiment is necessary. For cord blood haemopoietic stem cells, that were to be warmed rapidly, the optimum was cooling at 1–2°C min – with accelerated cooling below –60°C if desired [18]. Cooling rate is frequently an important determinant of recovery and it may be worthwhile to experiment with cooling rates above and below 1°C/min, say 0.3 and 3.0°C/min to be sure that recovery cannot be substantially improved by such a variation. If either variation produces an improvement then the trend should be followed until the actual optimum rate is located.

Two modes of controlled cooling have been advocated – actively controlled systems (programme-controlled cooling apparatus) and passive systems that rely on empirically optimised insulation to control cooling (the "plastic-box-in-a-minus–80°C-deep-freeze" approach). There are many other minor variations in method. It is not possible to choose between cryopreservation schemes without a rigorous comparison that takes due account of all the elements discussed above, including an appropriate choice of recovery assay, statistical design and method of analysis. As far as the cooling method goes, there is no reason why a properly optimised passive cooling system should not work well and studies have been published that fail to show a significant difference between the two cooling methods. However there also appear to be none that demonstrate the greater effectiveness of either method, or even that there really is no difference between them at an acceptable level of probability. Fundamentally, it is the cooling trajectory that matters, not the way in which it is produced. However, reproducibility, consistency, data recording and quality control also have to be considered and probably favour active programmed systems.

The storage temperature should be below –80°C if storage periods exceeding 1 year are envisaged but recovery is unlikely to be improved by using temperatures below –130°C when DMSO is the CPA. To provide a margin of safety, a storage temperature of –180°C is recommended. Immersion in liquid nitrogen is not necessary and should be avoided for reasons of safety. The optimum rate of warming during thawing is dependent upon the cooling rate that was used during freezing; if the cooling rate was optimised with thawing in a 37°C water bath then that technique of thawing should be retained.

Step 5: is to review the results. If they are not satisfactory, then an alternative cryoprotectant can be considered: glycerol and propylene glycol are perhaps the most obvious choices. Alternatively the effect of adding a non-penetrating solute such as albumin or HES to a lower concentration of CPA can be considered. In the case of cord blood stem cells neither is necessary to secure high recovery rates.

Cryopreservation of Multicellular Systems

Cell survival is essential for the preservation for surgical use of all organs and of many tissues. The standard method of preserving kidneys is by perfusion with a solution that mimics the composition of the intracellular fluid: disturbance of electrolyte balance across the cell membrane and increased water content during storage at +2 to +4°Care prevented. These solutions have a high potassium content and sufficient impermeant solute to prevent cell swelling when the active membrane pumps are turned down by cooling. There is still some debate about whether the initial flush with the preservation solution, followed by storage at 2–4°C, is sufficient or whether the organ should be continuously perfused throughout the storage period [19]. This was a popular topic for discussion 40 years ago but, has recently resurfaced [20] with increased commercial interest in the necessary hardware. The evidence does suggest that continuous perfusion can be beneficial, particularly when the organ was not of the very best quality in the first place.

The application of standard cryopreservation methods to organs, particularly to kidneys, has been studied – typically by perfusing with a solution containing ~10% DMSO or glycerol, cooling at 1°C/min, storing below –80°C, warming quickly and washing out the CPA. This approach has been uniformly unsuccessful with every vascularised organ that has been studied [21], with the possible exception of the ovary [22]. Conventional cryopreservation approaches have also been unsuccessful for corneas and articular cartilage, both of which require the survival of living donor cells in the graft.

Why then can so many isolated cells, but few tissues and no organs, be cryopreserved with functioning cells? When a cell suspension is cooled below its freezing point, ice crystals form and this can have two types of effect on the cells. If the suspension is cooled slowly, the ice forms first in the fluid around the cells and the cells experience only the change in composition of the solution that is caused by the removal of water. It is this indirect effect – the progressive concentration of solutes as freezing proceeds and the consequent cell shrinkage – that can be moderated by adding cryoprotectants. But if the system is cooled more rapidly, then water will not leave the cells sufficiently rapidly to avoid intracellular freezing: intracellular ice is very damaging. Tissues are obviously more complex than cell suspensions. For example, there may be problems of diffusion of solutes and heat through the mass of a tissue graft or there may be problems due to the fact there are different sorts of cell present in the tissue. Problems of the diffusion of solutes and heat are often discussed but in practice they are not crucial problems. A little more time may be required to add and remove the cryoprotectants and there may be a restriction of the range of cooling and warming rates that are physically possible but in practice these are not the critical problems in tissue cryopreservation. Similarly, because the curves of survival versus cooling rate flatten out at high cryoprotectant concentrations, cellular heterogeneity is less of a problem that might be predicted. Figure 8.1 shows some data for glycerol and red blood cells: when no cryoprotectant is present the cooling rate and the maximum cell survival are quite low, but as the cryoprotectant

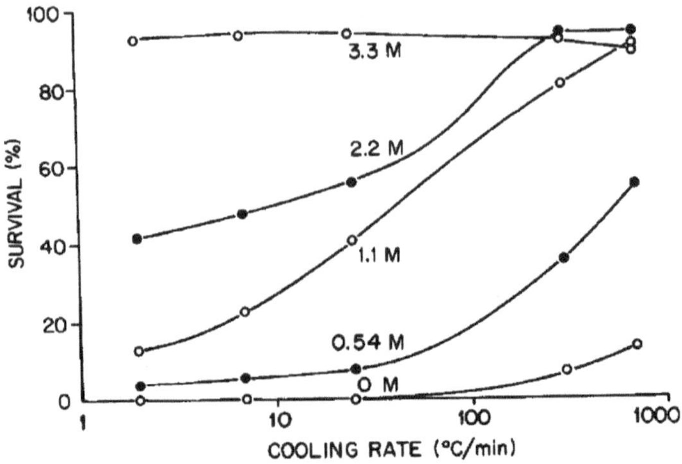

Fig. 8.1 The effect of cooling rate and glycerol concentration on the survival of human red blood cells following cryopreservation. Raising the glycerol concentration increases survival, especially at low cooling rates. With 3.3 M glycerol, survival is ~90% and is unaffected by cooling rate within the range tested

concentration is increased so more cells survive and the curve flattens [23]. If a sufficient concentration of cryoprotectant can be used, a cooling rate can be found that gives a high survival with different types of cell.

Although the presence of extracellular ice may not be a problem with cell suspensions, it can be a major problem with tissues. Figure 8.2 shows two micrographs of smooth muscle tissue: one had been cooled to –21°C at 2°C/min and the ice crystals were widely distributed throughout the tissue, including inside the muscle bundles; the other was cooled at 0.3°C/min and the ice formed predominantly in the

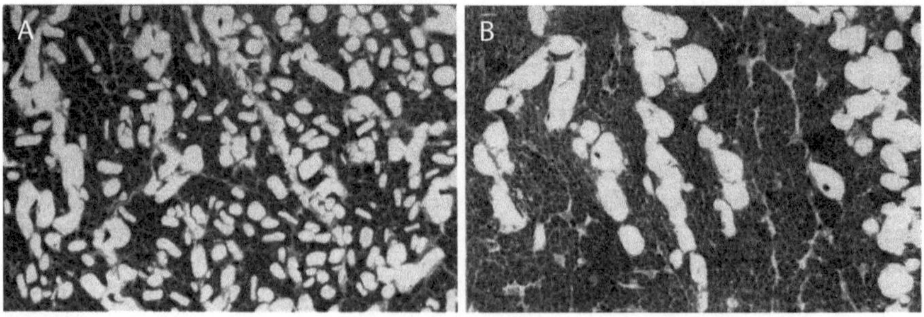

Fig. 8.2 Smooth muscle tissue, equilibrated with 20% DMSO, was cooled to –21°C at 2.0°C/min in (**a**) or at 0.3°C/min in (**b**) and then fixed at that temperature to preserve the ice morphology. At the higher cooling rate the ice was quite uniformly distributed throughout the tissue, but at the lower cooling rate the ice was predominantly restricted to the connective tissue between the muscle bundles. Greater function was retained in (**b**) than in (**a**)

connective tissue separating the muscle bundles. After cooling at 2°C/min the subsequent contractile function of that tissue was only 40% of control. At 0.3°C/min however, the contractile function was 60% of control [24, 25], demonstrating a correlation between the location of ice in the muscle bundles and poor function.

Figure 8.3 is a micrograph of articular cartilage that had been cryopreserved by a standard method: the white spaces contained the ice that had formed during cooling to –70°C. Most of the ice actually formed where the cells are, even though this tissue was cooled under conditions that enable a suspension of the same cells to survive. Chondrocytes that are frozen in situ in cartilage do not survive and the appearances show that much, if not all, the ice formed within the chondrocytes rather than the surrounding matrix. See Fig. 8.4. We know that intracellular freezing in cell

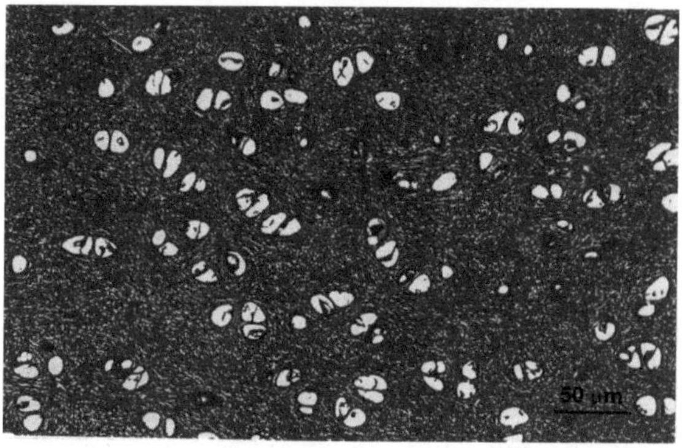

Fig. 8.3 The appearance of ovine articular cartilage that had been subjected to a standard DMSO/slow cool cryopreservation procedure and then freeze-substituted at –70°C. Note that large ice crystals, seen as clear spaces, are located in the chondrons. The extracellular matrix contains only finely granular ice crystals

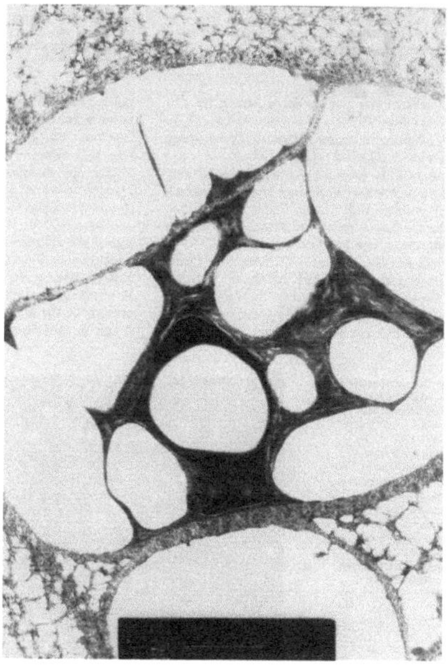

Fig. 8.4 An electrion micrograph of one chondron from cartilage that had been cryopreserved in the standard way that was used for Figure 3. Intracellular freezing had occured

suspensions is very damaging so it is not surprising that the chondrocytes in situ are killed, even when they are cooled slowly. The reason for intracellular freezing is not clear, but it may be that ice nucleates more readily in the chondrons than it does in the matrix.

Cryopreservation by Vitrification

One potential solution to this problem is to "vitrify" the system, which means avoiding the formation of ice by using such a high cryoprotectant concentration that the system becomes so viscous during cooling that it forms a glass but does not crystallize as ice. Vitrification avoids the formation of ice but it requires very high and therefore potentially toxic cryoprotectant concentrations. The concentration can be reduced by using very rapid cooling and warming because this inhibits the formation of ice crystals and this approach has been used very successfully with small volumes of cell suspension [26] but in the case of tissues the rates of change of temperature that are possible are restricted by the physical circumstances – relatively bulky samples. The result is that very high concentrations of cryoprotectant cannot be avoided and CPA toxicity is always the limiting factor.

We have been looking at a different approach to vitrification; although it is not really new. In 1965 John Farrant [27] carried out an experiment to provide evidence that the rise in salt concentration during the freezing of smooth muscle tissue was actually responsible for the observed freezing damage. To do this, he increased the concentration of cryoprotectant progressively as he cooled, following the freezing point depression curve for DMSO/saline. The samples were cooled until the temperature approached the liquidus line, which is the freezing point for that concentration of DMSO, and then the DMSO concentration was increased. The sample was then cooled again to just above the liquidus, and so on to –70°C. See Fig. 8.5. In this

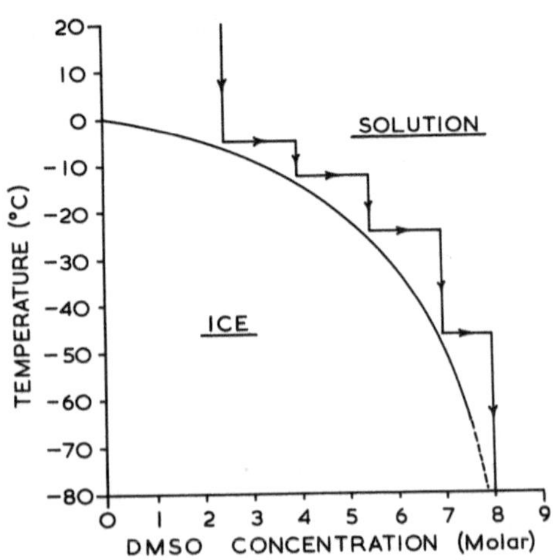

Fig. 8.5 Freezing point depression curve for DMSO/water with steps illustrating the method used to avoid ice formation without using very high initial DMSO concentrations

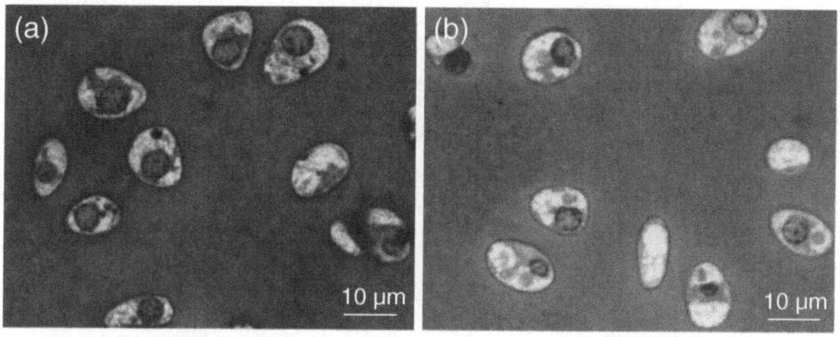

Fig. 8.6 Light micrographs of chondrocytes in situ in cartilage (**a**) Fresh tissue. (**b**) Tissue that had been processed by the liquidus tracking method such that no ice formed

way he obtained a very high final concentration of DMSO without a rise in salt concentration but also without ice forming. The approach was developed further by Elford and Walter in 1972 [28]. We have used this neglected approach, with some modifications, for cartilage [29, 30]. Figure 8.6a shows the appearance of chondrocytes in fresh cartilage. Figure 8.6b shows cells that had been through the whole preservation process – cooling down to –196°C (at which temperature they were vitrified) and then returning to room temperature. They appear normal. We also measured the ability of the cells to synthesise new glycoseaminoglycans (GAGs): this is an appropriate test of survival because that function is required in the cartilage graft and they possessed 70% of control ability to produce GAGs. Recent improvements in this method have raised the recovert of GAG synthesis to 87% of control. It is hoped that this approach will turn out to be applicable to other recalcitrant tissues.

Conclusions

The design of preservation methods for tissues rests upon the answer to the question "What properties is this tissue required to exhibit after it has been grafted?" For some tissues it will be intact cell structure and full metabolic function (organs; cornea;cartilage): for others, reproductive integrity (heamopoietic stem cells); or structure and full metabolic function (articular cartilage); or intact extracellular structure (tendon); or mechanical and biochemical integrity (bone). In some cases, incontrovertible clinical evidence for a particular set of functional criteria is not yet available but we do now have a wide range of preservation methods at our disposal and we do know what questions to ask.

Acknowledgments This chapter draws on material previously published by the author in the BBTS Newsletter No 42 (Autumn 1996) and in an article "The preservation of tissues for transplantation." Published in Cell and Tissue Banking, 7, 349–358 (2006).

References

1. O'Brien et al (1987) The viable cryopreserved allograft aortic valve. J Cardiac Surg 1(Suppl):153–167
2. Polge C, Smith AU and Parkes AS (1949) Revival of spermatozoa after vitrification and dehydration at low temperatures. Nature 164:666
3. Pegg DE, Wusteman MC, Boylan S (1997) Fractures in cryopreserved elastic arteries. Cryobiology 34:183–192
4. Pegg DE, Wusteman MC, Wang, L (2006) Cryopreservation of articular cartilage Part 1. Conventional cryopreservation methods. Cryobiology 52:335–346
5. Pegg DE (1989) Viability assays for preserved cells, tissues and organs. Cryobiology 26: 212–231
6. Huang Q, Pegg DE, Kearney JN (2004) Banking of non-viable skin allografts using high concentrations of glycerol or propylene glycol. Cell Tissue Bank 5:3–21
7. Hopkins RA (1989) Rationale for use of cryopreserved allograft tissues for cardiac reconstruction. In: Hopkins RA (ed) Cardiac reconstructions with aortic valves. Springer, New York, NY, pp 15–20
8. Aidulis D, Pegg DE, Hunt CJ, Goffin YA, Vanderkelen A, van Hoeck B, Santiago T, Ramos T, Gruys E, Voorhout W (2002) Processing of ovine cardiac valve allografts: 1. Effects of preservation method on structure and mechanical properties. Cell Tissue Bank 3:79–89
9. Neves J, Abecassis M, Santiago T, Ramos T, Melo J, Gruys E, Hulskamp-Koch C, Ultee A, Verkaar ELC, Lenstra CH, Goffin YA, Vanderkelen A, van Hoeck B, Hunt CJ, Pegg DE (2002) Processing of ovine cardiac valve allografts: 3. Implantation following antimicrobial treatment and preservation. Cell Tissue Bank 3:105–119
10. Barnes DWH, Loutit JF (1955) The radiation recovery factor: preservation by the Polge-Smith-Parkes technique. J Nat Cancer Inst 15:901
11. Lovelock JE, Bishop MWH (1959) Prevention of freezing damage to living cells by dimethyl sulphoxide. Nature(Lond) 183:1394–1395
12. Lovelock JE (1954) The protective action by neutral solutes against haemolysis by freezing and thawing. Biochem J 56:265–270
13. Ashwood-Smith MJ (1961) Preservation of mouse bone marrow at –79°C with dimethyl sulphoxide. Nature(Lond) 190:1204
14. Pegg DE, Lancaster PA (1998) A digital device and software for capturing and analysing cell volume data from a Coulter counter. Cryobiology 37:441
15. Arnaud FG, Pegg DE (1990) Permeation of glycerol and propane-1,2-diol into human platelets. Cryobiology 27:107–118
16. Hunt CJ, Armitage SE, Pegg DE (2003) Cryopreservation of umbilical cord blood: 1. Osmotically inactive volume, hydraulic conductivity and permeability of $CD34^+$ cells to dimethyl sulphoxide. Cryobiology 46:61–75
17. Pegg DE (2006) Principles of cryobiopreservation. In: Day JG, McLellan MR (eds) Cryopreservation and freeze-drying protocols. Methods in molecular biology, vol 38, 2nd edn. Human Press, Totowa, NJ
18. Hunt CJ, Armitage SE, Pegg DE (2003) Cryopreservation of umbilical cord blood: 2. Tolerance of $CD34^+$ cells to multimolar dimethyl sulphoxide and the effect of cooling rate on recovery after freezing and thawing. Cryobiology 46:76–87
19. Karow AM, Pegg DE (eds) (1981) Organ preservation for transplantation, 2nd edn. Marcel Dekker, New York, NY
20. Fuller BJ, Lee CY (2007) Hypothermic perfusion preservation: the future of organ preservation revisited. Cryobiology 54:129–145
21. Jacobsen IA, Pegg DE (1984) Cryopreservation of organs: a review. Cryobiology 21:377–384
22. Wang X et al (2002) Fertility after intact ovary transplantation. Nature 425:385
23. Morris GJ, Farrant J (1972) Interactions of cooling rate and protective additive on the survival of washed human erythrocytes frozen to –196°C. Cryobiology 9:173

24. Hunt CJ, Taylor MJ, Pegg DE (1982) Freeze-substitution and isothermal freeze-fixation studies to elucidate the pattern of ice formation in smooth muscle at 252 K (–21°C). J Microsc 125:177–186
25. Taylor MJ, Pegg DE (1983) The effect of ice formation on the function of smooth muscle tissue stored at –21° or –60°C. Cryobiology 20:36–40
26. Rall WF, Fahy GM (1985) Ice-free cryopreservation of mouse embryos at –196°Cv by vitrification. Nature (London) 313:573–575
27. Farrant J (1965) Mechanism of cell damage during freezing and thawing and its prevention. Nature (London) 205:1284–1287
28. Elford BC, Walter CA (1972) Effects of electrolyte composition and pH on the structure and function of smooth muscle cooled to –79°C in unfrozen media. Cryobiology 9:82–100
29. Pegg DE, Wang L, Vaughan, D (2006) Cryopreservation of articular cartilage 3. The liquidus-tracking method. Cryobiology 52:360–368
30. Wang L, Pegg DE, Lorrison J, Vaughan D, Rooney P (2007) Further work on the cryopreservation of articular cartilage with particular reference to the liquidus-tracking method. Cryobiology 55:138–147

Chapter 9
Sterilisation by Irradiation

Artur Kaminski, Izabela Uhrynowska-Tyszkiewicz, and Waclaw Stachowicz

Introduction

Sterilisation refers to any process that eliminates/inactivates transmissible infectious agents (pathogens) containing nucleic acids e.g. vegetative and spore forms of bacteria and fungi, parasites, viruses etc. There are two main types of sterilisation distinguishing by the characteristics of sterilisation agent used, namely physical and chemical sterilisation [3, 19].

Radiation sterilisation is a physical process based on the action of radiation energy with the matter. For sterilisation purposes ionising radiation is applied. The term *ionising radiation* covers all types of radiations carrying the energies capable to produce cascades of ionisations (ions formation) in the matter. Two types of ionising radiation are used for sterilisation purpose: electromagnetic rays with wave length below 100 nm (0.5/5.0 MeV) e.g. gamma and X-rays and corpuscular radiation i.e. fast moving mono-energetic electrons (3.0/10.0 MeV). Sterilisation efficacy of ionising radiation lies in its good penetrability in the matter followed by killing effect on pathogens [2, 6].

Both types of high energy radiations penetrate into the matter resulting in induction of ionisation of constituent molecules. However, both represent two different types of ionisation: direct or indirect.

Electrons possess both mass and charge. Therefore, they interact directly (direct ionisation) by electrostatic interaction with atoms which are close to their tracks which results in ionisation and formation of secondary electrons. These secondary electrons, in turn, are capable of initiating further processes of a similar character. Each interaction decreases the energy of fast electron and after several interactions its kinetic energy will be lowered and velocity reduced. At a defined depth electrons became thermalised and their activity is spent. It means that there will be

A. Kaminski (✉)
Department of Transplantology and Central Tissue Bank, Medical University of Warsaw, Warsaw 02-004, Poland
e-mail: akamin@ib.amwaw.edu.pl

G. Galea (ed.), *Essentials of Tissue Banking*, DOI 10.1007/978-90-481-9142-0_9,

no ionisation events in deeper layer of the material. Therefore, in contrast to photons, the range of penetration of high energy electrons is limited and depends on the initial energy of incident electrons, atomic number of absorbing material and its density. Moreover when high energy electrons pass close to atomic nuclei, they lose a considerable portion of their energy liberated in the form of photon radiation called *bremstrahlung* (German originated term for decelerating radiation). The higher atomic number of material, the higher is the decelerating radiation. In soft tissues the stopping of electrons is negligible, a little bit higher in bone tissue. In both cases it does not influence the course of radiation sterilisation.

The ionising radiation can induce not only ionisation events in the matter but also other physical and chemical processes. It happens if the energy absorbed raises an electron in an atom/molecule to a higher energy state without ejection of electron, breaking of chemical bounds or inducing nuclear reaction. It is possible to distinguish three consecutive stages of the interaction of ionising radiation with the matter: (1) physical stage – excitation and/or ionisation of its atoms/molecules. (2) physicochemical stage – free radicals formation and (3) chemical stage – reactions between free radicals and free radicals with atoms/molecules of the matter [23]. In biological systems as a result of radiation induced physical, physicochemical and chemical processes some changes of biological properties are observed (biological stage). The appearance of high chemically reactive particles such as free radicals, ion-radicals etc. may result in changes on the molecular level which in turn lead to changes on the cellular level. For example, destruction of proteins by the mechanism of peptide links breaking, amino group NH_2 removal (deamination), disulfide bridges formation etc. leads to changes in proteins conformation and to decline the function either of enzymes and of structural proteins. Changes in lipids such as lipid peroxidation may result in changes in the permeability of cellular membranes and may lead to release of lysosoms contents, changes in processes of protein synthesis in rough endoplasmic reticulum (RER), formation of toxic substances and disturbance of mitochondrial oxidative phosphorylation. Apart from biological membranes the DNA is primary radiobiological target. Irradiation-induced DNA lesions may consist in breaks in DNA strands, alteration to bases (their oxidation, alkylation, hydrolysis, adduct formation), destruction of sugars as well as crosslinks and dimmers formation. All these varied and interrelated changes occurring at the biological stage are formed within seconds, hours, days, and throughout the life of the cell and can lead either to the premature cell death by the mechanism of apoptosis or oncosis, or, in the case of nonlethal lesions, be passed on from parent cells to their progeny [25].

It was estimated that a secondary 1 MeV electron passing through the liquid medium produces about 40,000 ionisations and about the same number of excitations. Ionised molecules are neutralised in a very short time (10^{-11} s) in reaction with low energy electrons appearing in excess during radiation treatment and become highly excited parent molecules. Most of the excited molecules lose the excess energy in contact with surrounding molecules while a small fraction of them decompose into free radicals which may undergo further transformation in reaction with radicals, atoms and molecules collocated in their neighbourhood. This is a typical way of the formation of radiolytic products in the condensed phase. In crystalline matrices radiation induced ions and radicals may survive in deep traps for a longer period.

The different mechanisms of the absorption of gamma rays and fast electrons has important practical implications.

Specificity of Radiation Sterilisation with Gamma Rays and Electron Beams

Gamma and X rays reveal much better penetrability inside the irradiated material as compared with fast electrons. In consequence, more uniform distribution of dose inside the graft is expected under gamma exposure. Instead, electron beams are more effective in producing ionisation per mass unit when compared with gamma photons. It means that to reach the same level of sterility gamma radiation has to be applied much longer (several hours) than electron beams (several minutes).

On the other hand, penetrability of gamma rays is higher than electrons. Consideration on the problem of the penetration and homogeneity of ionizing radiation in the bulk of grafts relies on the assumption that the average density of the grafts is 2 g/cm^3 the number better fitted to the real value than 1 g/cm^3 (density of water) typically taken earlier in this type of consideration.

The reduction of the dose of gamma rays by 50% occurs in 2 g/cm^3 model system at the depth of 6 cm while the same effect is obtained at the depth of 1.8 cm only if the beam of electrons is applied. Consequently, at the depth of 2.6 cm electron beam (EB) irradiation is practically reduced to zero level. Such situation never appears with gamma rays which are absorbed exponentially (Fig. 9.1). The specific feature of ionising radiation is a dose build-up inside irradiated objects. It is effected

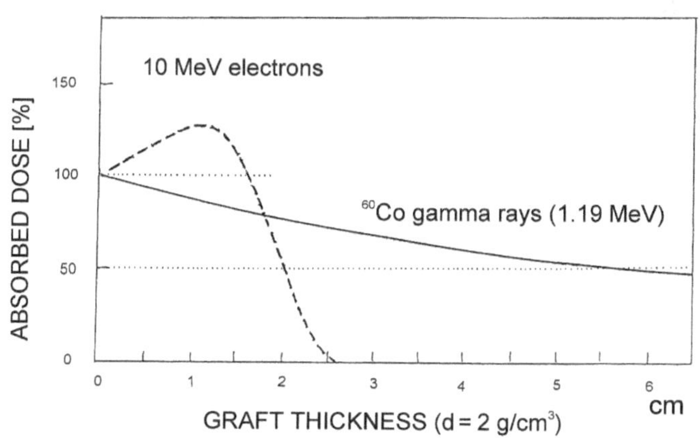

Fig. 9.1 Dose depth distribution of ^{60}Co gamma rays and beams of 10 MeV electrons in a model matrix of the density 2 g/cm^3. One-side irradiation. *Dotted line* – 10 MeV electrons. *Continuous line* – ^{60}Co gamma rays. From: Stachowicz [22]

by a violent ionisation in the upper inside layer of irradiated product not followed by neutralization process fast enough. In deeper layers of the product effected by proceeding thermalisation, the excess electrons are neutralised fast and the dose decreases as a function of absorbed radiation energy. The effect is pronounced with electron beam irradiation. As a consequence of this effect at the depth of 1.3 cm the dose of 10 MeV electron beam is by 27% higher than in the surface layer (Fig. 9.1).

In order to improve the homogeneity of irradiation dose inside the graft it is advisable to apply two-side radiation treatment. With a two-side gamma irradiation the acceptable depth-dose distribution is obtained inside the graft 15 cm thick (Fig. 9.2), while with two-side EB irradiation, the thickness of ca 4.7 cm will meet the requirement (Fig. 9.3c). Four graphs presented in Fig. 9.3 show evidently how much the distribution of dose inside the graft depends on its thickness [13]. With the matrix only slightly thicker (5.2 cm) the dose in the middle of a graft will be lower than accepted for safe surgery (Fig. 9.3d). On the other side, with a graft of lower thickness (3.7 cm) the dose inside the matrix will be enormously higher then exposure (surface) dose. This would not be acceptable from the point of view of mechanical resistance of the implant (Fig. 9.3a). In this case an acceptable decrease of the exposure dose by ca 30% could be acceptable solution. By sterilisation of "light grafts" (skin, cartilage etc) or thin bone bars, the EB irradiation from one side could be satisfactorily enough. The decision whether to adapt one-side or two-side EB treatment or whether it is necessary to modify sterilisation process in some detail depends on the operator of irradiation facility [22].

Radiation sterilisation is classified as a "cold" process what means that in the course of irradiation the temperature is not changed significantly and does not influence the process. This is generally true but one needs to remember that, the last

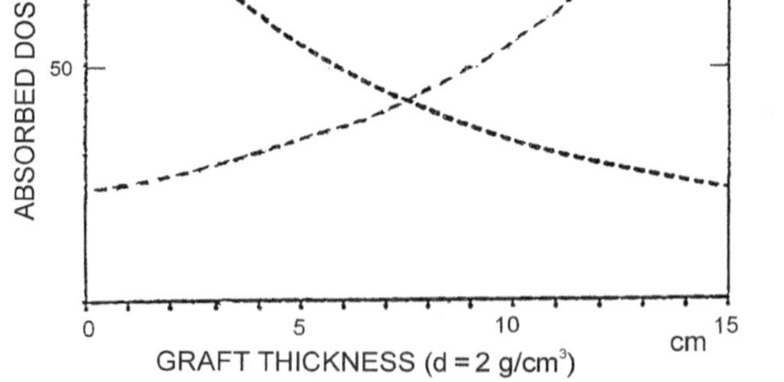

Fig. 9.2 Dose depth distribution of ^{60}Co gamma rays in a model matrix of the density 2 g/cm^3. Two-side irradiation. Thickness of irradiated model matrix 15 cm. From: Stachowicz [22]

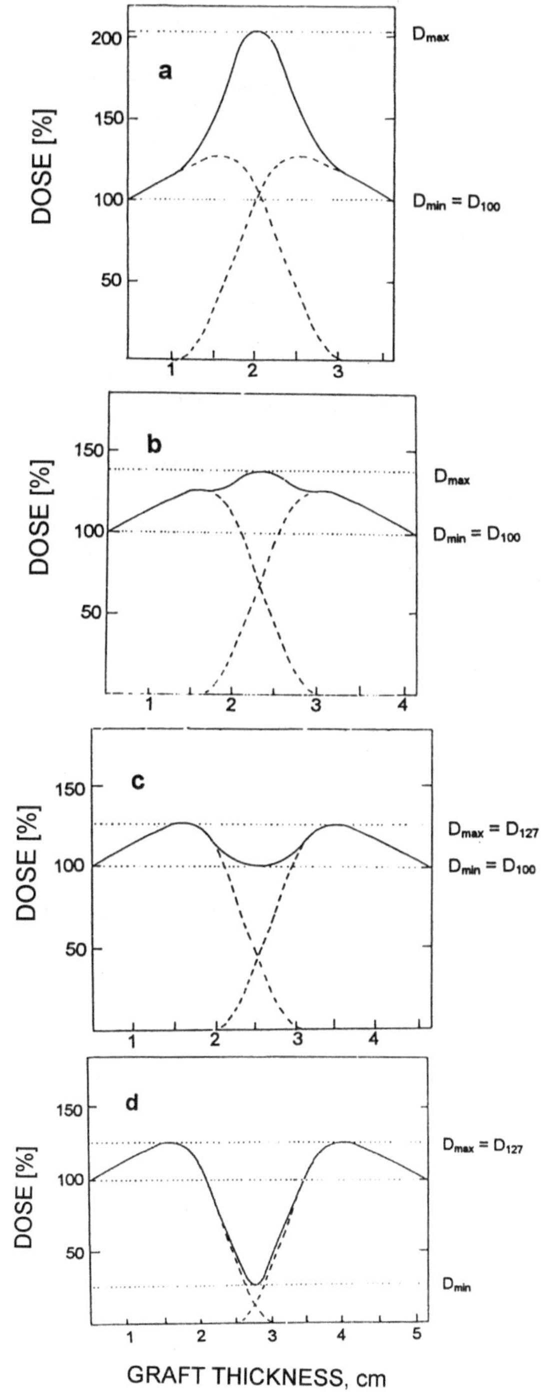

Fig. 9.3 Dose depth distribution of a beam of 10 MeV electrons in a model matrix of the density 2 g/cm^3. Two-side irradiation. (**a**) thickness 3.6 cm: the dose in the centre is two times higher than the surface dose; (**b**) thickness 4.2 cm: the dose in the centre is by 50% higher then the surface dose; (**c**) thickness 4,7 cm: the dose inside the matrix is by 27% higher then the surface dose; (**d**) thickness 5,2 cm: the dose in the centre is by 75% lower then the surface dose. Graph **a** illustrates the over-dosage in the centre of the graft, while graph **d** the case with a dose in the centre which is too low to be accepted. From: Stachowicz [22]

step of the absorption of ionizing radiation described earlier as electron thermalisation process is followed by local emission of heat inside the graft. The temperature rise of tissue allografts irradiated with the dose of 35 kGy of gamma rays does not exceed a few degrees. Since EB irradiation is more effective then gamma irradiation, the thermalisation process inside the EB treated graft becomes more effective too and more heat is locally emitted in a unit of time. Therefore, it is advisable to conduct EB sterilisation (25/35 kGy) of grafts in two steps to deliver the energy in two smaller portions.

Tissue grafts prepared for radiation sterilisation must be isolated from the outer atmosphere. Usually they are sealed in double or triple bags made of polymeric or laminated foils 0.02–0.05 mm thick. A foil must be resistant against higher doses of radiation and non-reactive against organic substances, like fatty acids, for example, that may be present in the tissue graft. Polycarbonates, polystyrene, polyesters, polyethylene and their laminates are commonly in use. It is recommended to apply commercial materials specifically produced for this purpose. For radiation sterilisation with gamma rays hard, thick-wall, plastic or glass containers are also acceptable. It has to be remembered, however, that some not significant part of the energy of ionizing radiation is absorbed in such package [9–11].

Dosimetry for Radiation Sterilisation

Absorption of ionizing radiation is non-homogenous in its nature and special attention has to be paid to a proper modelling of irradiation procedure. It is obvious that any piece of bone prepared for grafting is not of the same density and that each graft differs from the other in its shape and local densities. This implies variations of the absorbed irradiation energy i.e. variations of dose absorbed inside each of graft. It is why dosimetry problems are more complex with allografts then with medical materials sterilized by radiation but manufactured with the use of plastics. It has to be remembered that control of dose is an essential problem for radiation sterilisation of tissue grafts [15].

The absorbed dose *D* (simply called dose) is the amount of the energy absorbed per unit mass of irradiated product at the unit point in the region of interest. The SI unit of dose is 1 gray (Gy) corresponding to 1 J of energy absorbed in 1 kg of a product. In a practical situation, *D* is given as an average value, however, the upper and lower limits of dose delivered to the product should be formulated too. Any way, the absorbed dose or "sterilisation dose" the term applied commonly to tissue grafts is in fact the average dose only. It is practically the exposure (surface) dose optimised from the point of view of microbial and structural (mechanical and biological properties) demands. However, in order to assure desired sterility, more detailed information on the distribution of dose inside the graft are needed.

The adapted exposure dose must be calibrated with reference dosimeters. In addition, the individual dose measurement on each tissue graft must be done. For this purpose dose-meters attached to every package or bag with a graft are used. The following are commonly used:

(i) A sensitive water or graphite calorimeter adapted for the estimation of absorbed dose and dose rate of electron beam (EB) irradiation. It is based on electrometric measurements of the increase of the temperature during exposure;
(ii) Ferrous sulfate solution (Fricke dose-meter) for the estimation of absorbed dose and dose rate during gamma-ray irradiation. The concentration of ferric ions produced by radiation oxidation of Fe^{2+} ions are measured spectrophotometrically [8].

Both dosimeters should be related to international standards. It is achieved by the use of transfer dosimeters controlled by authorised international dosimetry center.

The important function of routine dosimeters (plastic films, solutions, powders, glasses) is to monitor package-to-package or bag-to-bag dose variations and to facilitate the evaluation of dose distribution in tissue grafts.

The detail information on dose distribution through the product is in practice not as important as the determination of two extremes:

(i) a maximum dose D_{max} absorbed in the graft which must be lower than or equal to the highest acceptable dose, and
(ii) a minimum dose D_{min} absorbed in the graft which, in turn, must be greater than or equal to the lowest acceptable dose.

By sterilisation of allografts the most important is the knowledge of D_{min} since this number corresponds to the lowest dose in some part of a graft in the which the sterility must be assured to guarantee the sterility of the latter in total (see also Fig. 9.3).

D_{max} must be always lower than critical level of dose accepted for a given type of a graft. The methods for the determination of both minimal and maximal doses with the use of routine dosimeters will be discussed.

From a fairly large number of routine dosimeters presently applied in irradiation facilities, the most suitable for dose measurements on tissue grafts seem transparent plastic dosimeters distributed in the form of blocks or tapes adapted for the use in spectrophotometry. The dose estimation is based on spectrophotometric measurement of absorbency (optical density) which is proportional to the dose applied. The measurement of absorbency at a suitable wavelength, usually in near ultraviolet allows to determine the dose with a reasonable accuracy.

Plastic dosimeter most frequently used is based on poly-methylmethacrylate (PMMA) and permits to measure the dose from 0.1 to 50 kGy. It is usually available in small blocks 2–3 mm thick. There are several types of PMMA dosimeters in use which absorb at different wavelengths. A family of plastic film dosimeters from 0.01 to 0.20 mm thick includes radiochromic film dosimeters containing dyes sensitive to radiation or pure polymer dosimeters changing their colouring during the irradiation. Poly-chlorostyrene films with the content of triphenylmethane dye, polyethylene tetraphtalate or pure polyvinyl films are most frequently used.

Polyvinyl chloride (PVC) dosimeter is relatively cheap and enables to measure the dose at 398 nm within the range 5÷50 kGy. It can be also used in the form of long tapes to be collocated on the packages with graft lengthwise the irradiation container to control the homogeneity of electron beam and conveyer speed during EB irradiation. The PVC dosimetric films in cartoons are used for the determination of D_{min} if placed below a single bag with tissue graft in the button of a carrier. The measurement can be done with one-side irradiation to decide whether this method could be adapted for a given kind of graft. Thick grafts can be controlled by a "sandwich method". In order to adapt this technique it is necessary to perform a model experiment. A representative graft is cut in two equivalent parts and between these parts a sheet of dosimetric foil is collocated. Spectrophotometric examination gives the answer whether the graft of this type can be sterilized properly by EB irradiation. The procedure is called *dose mapping* (Fig. 9.4). When applied to commercial gamma irradiator with a multi-side exposure system, the distribution of tens of smaller dosimeters throughout one representative package with tissue grafts is needed. The examination of all dosimeters allows for the proper estimation of the extreme values of dose in the package. There is no need to adapt this procedure if single bags with tissue grafts are irradiated by gamma rays. L-alanine dosimeter and bone dosimeters are also in use [1, 16, 17]. Both polycrystalline l-alanine and bone powder give rise in electron paramagnetic resonance spectroscopy (epr) to stable and specific signals shown in Fig. 9.5. The amplitude of the central line (h_c) or the integrated area of the epr signal is proportional to absorbed dose of gamma or EB radiation. The dose-dependence curves of the response of both dosimeters are shown in Fig. 9.6. The advantage of bone powder dosimeter is that its composition is adequate with the composition of bone, the tissue most frequently used for

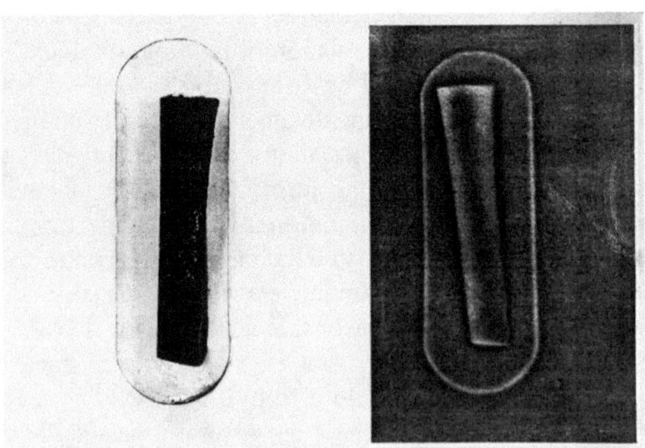

Fig. 9.4 PCV mapping of a small bone bar in a plastic container. One-side irradiation. On the *left* a photo, on the *right* mapping with PCV foil. The areas of different degree of darkening indicate different levels of absorption of ionizing radiation inside the graft. Deeper darkening (*gray*) corresponds to a suitable, relatively uniform absorption of ionising energy in the bulk of graft. *Light colour* indicates almost full absorption of ionising energy in the graft but followed by markedly lower absorption in this part of graft which is opposite to the surface exposed to radiation. From: Dziedzic-Goclawska and Stachowicz [25]

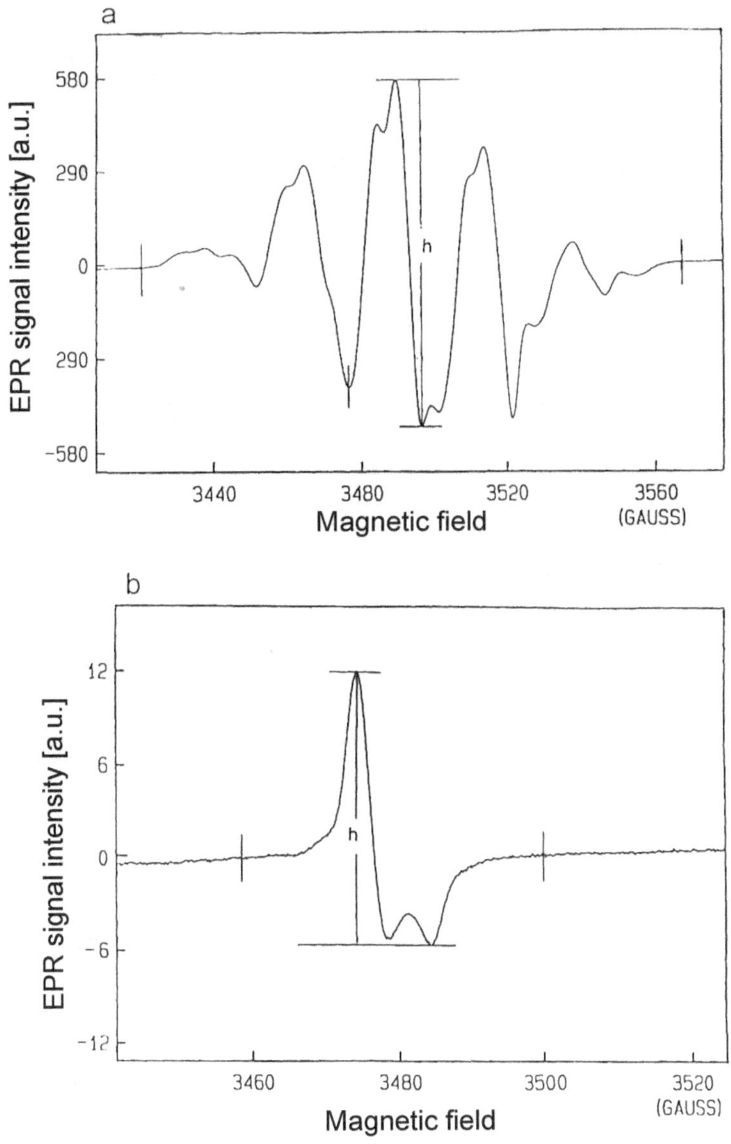

Fig. 9.5 Epr signals (first derivative) recorded with l-alanine (a) and bone (b) dosimeters. The estimated dose is calculated from the proportion of the heights (h) of the peaks. From: Dziedzic-Goclawska and Stachowicz [25]

grafting. Bone dosimeter is produced from bovine bone (*femur*) powder with average grain size of 0.02 mm. The preparation comprises special cleaning, sieving and lyophilization of the powder. Bone dosimeter can be used within the range of doses from 0.05 kGy to about 40 kGy of gamma rays or beams of accelerated electrons. The curvature in dose vs signal intensity relationship around 25 kGy makes the

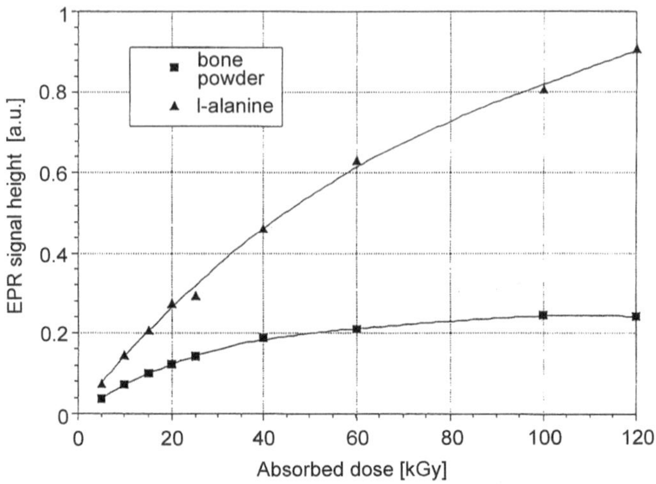

Fig. 9.6 Relationship between the heights of the epr signals recorded with l-alanine and bone dosimeter as a function of dose of ^{60}Co radiation. From: Ziaie et al. [24], 603–608

dose estimation at higher doses less accurate. It has been proved that the error of dose estimation below 25 kGy is from 3 to 4%, while above this level is increasing gradually to 8% around the dose of 40 kGy [20, 21, 24].

Routine dosimeters are calibrated with reference dosimeters described earlier and all of them are based on the readings from dose-dependence response curves or alignment charts. In the case of significant difference in the dosimeter and product densities the corrections are sometimes needed.

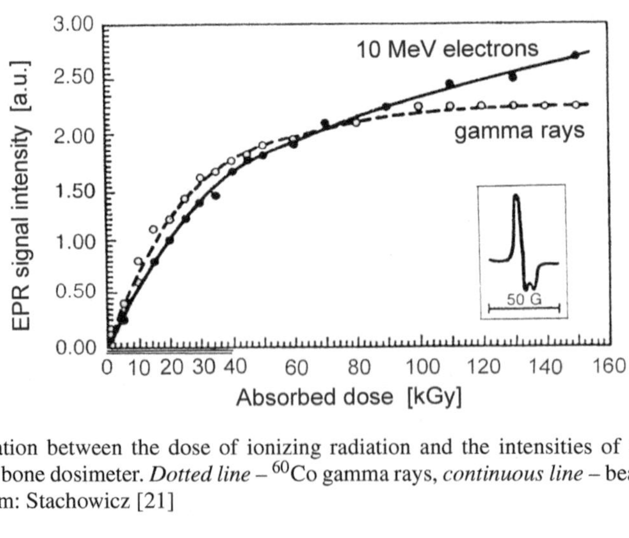

Fig. 9.7 Relation between the dose of ionizing radiation and the intensities of the epr signals recorded with bone dosimeter. *Dotted line* – ^{60}Co gamma rays, *continuous line* – beams of 10 MeV electrons. From: Stachowicz [21]

The response of reference and routine dosimeters to gamma rays and EB irradiations is not much different. Slight difference not exceeding a few percent has been reported. In addition to the described dosimeters to each bag with allograft so called "go-no-go" radiation-sensitive indicators are routinely attached. The role of indicators is to ascertain visually whether given graft was or was not irradiated. During radiation treatment indicators change their coloration from yellow to red, for example. It has to be stressed that dye indicators cannot be used for the measure of dose and for that reason can never be considered substitute for routine dosimeters.

Mechanisms Involved in the Inactivation of Micro-Organisms by Ionising Radiation

The major target sites in micro-organisms that are susceptible to ionising radiation are nucleic acids (DNA – deoxyribonucleic acid, RNA – ribonucleic acid). The damaging process may be caused directly by ionising radiation or indirectly through radiolysis of water and the production of highly reactive, short-lived hydroxyl radicals (*OH). In the presence of water, the indirect mechanism predominates. The presence of oxygen enhances the damaging effect. Oxygen reacting with hydroxyl radicals produces peroxide radicals and peroxides cause various kinds of damage to DNA. Both direct and indirect effects of ionising radiation may cause single or double strand breaks of DNA. DNA intra-strand cross-links and damage to the DNA bases or DNA sugar may also occur. Ionising radiation induces structural damage in DNA which inhibits DNA synthesis, causes errors in protein synthesis, and this leads to cell death. It has to be kept in mind that at low doses of radiation, several bacteria possess the ability to repair damage to DNA due to the action of repair enzymes (e.g. DNA polymerase I) and recombination. Repair of single-strand brakes and double-strand brakes (which are of course more difficult to repair) produce radiation-resistant mutants such as *Deinococcus radiodurans (Micrococcus radiodurans).*

Radiation Resistance of Micro-Organisms

The radiation resistance of micro-organisms is genetically determined. Gram-negative bacteria are more sensitive than Gram-positive bacteria. Usually, spores are more radiation resistant than the vegetative forms of bacteria. The most resistant fungi may be as resistant as bacterial spores, while viruses are, in general, more resistant than bacteria. Prions are extremely resistant to most chemical and physical sterilising agents, including ionising radiation.

Enzymes, pyrogens, toxins and antigens of microbial origin are, in general, very radiation-resistant, compared to living cells. Therefore, the number of micro-organisms present prior to radiation sterilisation is of importance when dealing with medical materials, regardless of the radiation resistance of the contaminating population.

There are many factors that can modify the radiation resistance of micro-organisms. Protectors such as alcohol, glycerol, reducing agents, dimethyl sulphoxide (DMSO), proteins and carbohydrates increase resistance. On the other hand, the presence of water and oxygen will enhance radiation damage. Irradiation at low temperatures increases, while at higher temperatures decreases, the resistance of micro-organisms. These factors should be taken into consideration when setting the sterilisation dose for any product.

An acceptable sterilisation procedure for any type of product, including tissue allografts, depends on defining the most resistant micro-organisms which could be present and the density of each of these.

The results of the radiation sterilisation procedure depend on the amount of energy transferred, the number of contaminating micro-organisms and their resistance to ionising radiation (initial contamination) characterised by the D_{10} values.

The commonly used term "bioburden" describes the population of viable (active) micro-organisms that are present on or inside a material or product before sterilisation. This is one of the factors influencing the efficiency of radiation. Clearly, the lower the bioburden is, the more effective the process occurs.

The D_{10} value, usually expressed in kGy, is the dose of irradiation necessary to reduce the initial microbial population by 1 $\log_{10}$, i.e. by 90%. This value can be read directly from the dose-inactivation curve or calculated using the following equation:

$$\text{D-value} = \frac{\text{radiation dose}}{\text{LogN}_0 - \text{logN}}$$

where N_0 and N represent 1–log difference in viable numbers. D-value differs greatly among different types of micro-organisms and, occasionally, considerable variation among different strains of the same organism is observed. The response of micro-organisms to radiation also depends on external conditions.

The concept of "sterility assurance level" – SAL is derived from kinetic studies on microbial inactivation, i.e. the probability of viable micro-organisms being present on or inside a product unit after sterilisation. For example, SAL 10^{-6} would assure that less than one out of a million contaminants would survive on or inside the product following sterilisation. Depending on the risk posed by the use of various specimens, different values of SAL (10^{-3}, 10^{-6}) may be recommended. For medical devices which are in contact with blood, for parenteral solutions as well as for tissue allografts, a value of SAL 10^{-6} or lower is required.

Radiation sterilisation of health care products, which was introduced in the middle of the 1950s, has recently been more frequently used. The sterilisation dose recommended for these products is 25 kGy, and this is the dose that is commonly used in the UK, in the USA and in many other countries. In some Scandinavian countries, the recommended dose is up to 45 kGy. The recommendation of a dose of

25 kGy is based upon the bioburden and radiation resistance of micro-organisms that are found on health care products. International standards specifying procedures for the validation and routine control of the process used for sterilisation of health care products have been prepared by the International Organization for Standardization.

The same dose of 25 kGy has been recommended for sterilisation of tissue allografts and is routinely applied in many tissue banks, Although this dose may not be entirely satisfactory, as explained below.

In cases of health care products that are manufactured under defined and clean conditions, it is easy to establish the average bioburden, which is usually low and has a standard distribution, thus the recommended dose of 25 kGy has been accepted for most of these products.

With respect to human tissues that are collected from cadaveric (or even from living) donors, it is very difficult or simply impossible to determine the bioburden each time, since initial contamination may vary greatly from tissue to tissue and from one donor to another.

The problem is additionally complicated by the possible presence, in human tissues, of pathogenic viruses such as the human immunodeficiency virus (HIV), hepatitis viruses (HBV, HCV), cytomegalovirus (CMV) or others. Data concerning the sensitivity of these viruses to ionising radiation are scarce. The effectiveness of ionising radiation to inactivate them in tissues that have been collected from cadaveric donors has not been well documented and the mechanisms of viral responses are unclear. This is mainly due to the fact that there are no suitable tests to study their inactivation, no appropriate animal model exists and no suitable method of in vitro culture of the highly differentiated target cells for these viruses has been developed.

In retrospective studies, transmission of hepatitis C virus (HCV) through non-sterilised cadaveric tissue allografts has been reported [4], while allografts irradiated with a dose of 17 kGy did not evoke infections in graft recipients. There are no data on the sensitivity of hepatitis B virus (HBV) to ionising radiation in tissues.

The majority of studies have been carried out on the inactivation of HIV. It has been postulated that the dose of irradiation needed to reduce the viral load by 1 $\log_{10}$ – the D_{10} value is 4 kGy or even 5.6 kGy. Taking into consideration the required SAL 10^{-6}, assuming the average HIV bioburden to be about 10^3 virions/mL for the state of acute infection and a D_{10} value of 4 kGy, a reduction of 9 (6 + 3) units or a dose of 36 kGy would be required. On the other hand, if the D_{10} value is 5.6 kGy, then a dose of > 50 kGy would be needed to inactivate HIV. These results are in agreement with data that has been published by Fideler who, using a polymerase chain reaction (PCR), found that in order to inactivate HIV in fresh frozen bone–patellar-ligament–bone grafts, a dose in the range of 30–40 kGy is required. The sensitivity of HIV to ionising radiation also depends on the temperature of irradiation. The reduction of the virus titre of 5–6 $\log_{10}$ was achieved at doses 50–100 kGy in frozen plasma (–80°C) and 25 kGy at 15°C [5, 7, 12].

Considering the high D_{10} value for HIV, even a dose of 35 kGy cannot be treated as the sterilisation dose, but it is impossible to increase the irradiation dose with impunity, as high doses of ionising radiation (over 50 kGy) evoke many physico-chemical changes in tissue allografts which may affect their biological properties.

The selection of a radiation dose is a compromise between a dose that is high enough to inactivate as many micro-organisms as possible and low enough to preserve important biological properties of tissue allografts [14, 18]. It is recommended to implement a dose of 35 kGy, which certainly provides a more adequate assurance of sterility for tissue allografts than does the commonly used dose of 25 kGy.

The validation of radiation sterilisation processing of tissue allografts should be performed by adequate measurements of the absorbed dose, set a priori and required to achieve the specified sterility assurance level (SAL). In addition to proper dosimetry systems, it is advisable to use radiation-sensitive indicators. The purpose of these "go-no-go" indicators is the check visually whether or not a graft has been irradiated. An indicator should be attached to every graft package and it should entirely change colour after irradiation. It should be stressed that indicators are not dosimeters and should never be used as a substitute for routine dosimeters.

Factors Affecting the Effectiveness of Radiation Sterilisation and the Sensitivity of Micro-Organisms to Irradiation

Several factors can affect the effectiveness of radiation sterilisation and can modify the microbial sensitivity to ionising radiation.

One of the factors influencing the effect of irradiation is bioburden, i.e. the number of micro-organisms present on or inside tissues before sterilisation.

Since toxins, pyrogens and antigens of bacterial origin are very resistant to irradiation, in comparison to living cells, regardless of the radiation resistance of the contaminating population, it is strongly recommended to avoid recontamination of tissues during retrieval and processing and to prevent micro-organism proliferation by temporary storage of tissues at low or freezing temperatures before sterilisation.

The presence or absence of water and oxygen plays an important role. In the absence of water (e.g. in air dry or lyophilised samples), the resistance of micro-organisms is increasing. On the other hand, in the presence of water, an indirect effect of ionising radiation predominates. Oxygen enhances the damaging effect to micro-organisms and increases their sensitivity to irradiation. During radiolysis of water, highly reactive, short-lived hydroxyl radicals (*OH) which react with oxygen are produced, causing the formation of peroxide radicals and peroxides which intensify damage to the nucleic acids of micro-organisms.

Therefore, if lyophilisation is used as a preservation procedure, it would be better to leave some water in the tissue than attempt to remove as much water as possible. It has also been recommended that the vials containing tissues be refilled with inert gaseous nitrogen to remove as much atmospheric oxygen as possible. However, since oxygen increases the sensitivity of micro-organisms to ionising radiation, this procedure should be discontinued.

It should be noted that irradiation at low temperatures increases, while at higher temperatures decreases, the resistance of bacteria and viruses. It has been found that the resistance further increases when lyophilised viruses are irradiated at a low (–79°C) temperature as compared to fresh-frozen irradiation at –79°C.

All these factors influence the effectiveness of radiation sterilisation and should be taken into account when setting the radiation dose.

References

1. Brady MN, Aarestad NO, Swartz HO (1968) In vivo dosimetry by electron spin resonance spectroscopy. Health Phys 15:43–47
2. Bailey AJ (1968). Effect of ionizing radiation on connective tissue components. In: Hall DA (ed) International review of connective tissue research, vol 4. Academic Press, New York, NY, 233 pp
3. Christensen EA, Kristensen H, Miller A (1992) Radiation sterilisation. A. Ionizing radiation. In: Russell AD, Hugo WB, Ayliffe GAJ (eds) Principles and practice of disinfection, preservation and sterilization, 2nd edn., Blackwell Science, Oxford, 528–543 pp
4. CDC (2003) Hepatitis C virus transmission from an antibody-negative organ and tissue donor – United States, 2000–2002. MMWR Morb Mortal Wkly Rep 52(13):273–276
5. Conway B, Tomford W (1992). Radiosensivity of human immunodeficiency virus type 1. Clin Infect Dis 14:978–979
6. Dziedzic-Goclawska A, Kaminski A, Uhrynowska-Tyszkiewicz I, Stachowicz W (2005) Irradiation as a safety procedure in tissue banking. Cell Tissue Bank 6(3):201–219
7. Fideler BM, Vangsness CT, Moore T, LI Z, Rasheed S (1994). Effects of gamma irradiation on the human immunodeficiency virus. J Bone Joint Surg 76A:1032–1035
8. Fricke H, Hart EJ (1966) Chemical dosimetry. Radiation Dosimetry, vol II, 2nd edn. Academic Press, New York, NY
9. Guideline for Electron Beam Sterilization of Medical Devices (1990) AAMI ST-31-066-SM
10. Guideline for Gamma Radiation Sterilization (1991) ANSI/AAMI ST 32
11. Guidelines for Industrial Radiation Sterilization of Disposable Medical Products (Cobalt-60 Gamma Irradiation) (1990) TECDOC – 539, January 1990
12. Ho DD, Moudgil T, Alam M (1989) Quantitation of human immunodeficiency virus type 1 in blood of infected persons. N Engl J Med 321:1621–1625
13. International Atomic Energy Agency (2002) Dosimetry for Food Irradiation, TRS (Technical Reports Series) no 40, IAEA Vienna
14. Komender J, Malczewska H, Komender A (1991) Therapeutic effect of transplantation of lyophilized and radiation-sterilized allogenetic bone. Clin Orthop 272:38–49
15. McLaughlin WL, Boyd AW, Chadwick KH, McDonald JC, Miller A (1989) Dosimetry for radiation processing. Taylor and Francis, New York, NY
16. Owczarczyk HB, Migdał W, Stachowicz W (2002) EB dose calibration for 10 MeV linear accelerator. Rad Phys Chem 63:803–805
17. Regulla DF, Deffner U (1982) Dosimetry of ESR spectroscopy of alanine. Appl Radiat Isot 33:1101–1109
18. Rock MG (1996) Biomechanics of allografts. In: Czitrom AA, Winkler H (eds) Orthopedic allograft surgery. Springer, New York, NY, pp 29–39
19. Russell AD (1999). Radiation sterilization. A. Ionizing radiation. In: Russell AD, Hugo WB, Ayliffe GAJ (eds) Principles and practice of disinfection, preservation and sterilization, 3rd edn. Blackwell Science, Oxford, pp 675–687
20. Stachowicz W, Michalik J, Dziedzic-Goclawska A, Ostrowski K (1972) Deproteinized bone powder as a dosimeter for radiosterilisation of biostatic grafts. Nukleonika 18:425–431
21. Stachowicz W, Michalik J, Dziedzic-Goclawska A, Ostrowski K (1972) Evaluation of absorbed dose of gamma and X-ray radiation using bone tissue as a dosimeter. Nukleonika 19:845–850

22. Stachowicz W (2004) Zagadnienia techniczne sterylizacji przeszczepów tkankowych za pomocą promieniowania gamma i szybkich elektronów. In: 40 lat Bankowania i Sterylizacji Radiacyjnej Tkanek w Polsce. Red.: A. Dziedzic – Gocławska i inn. Zakład Transplantologii i Centralny Bank Tkanek Akademii Medycznej w Warszawie, Instytut Chemii i Techniki Jądrowej, Warszawa, pp 95–104
23. Zagórski ZP (1981) Sterylizacja Radiacyjna (Radiation Sterilisation). In: Polish. PZWL, Warszawa
24. Ziaie F, Stachowicz W, Strzelczak G, Al-Osaimi S (1999) Using bone powder for dosimetric system. EPR response under the action of γ irradiation. Nukleonika 44(4):603–609
25. Dziedzic-Goclawska A, Stachowicz W (1997) Advances in tissue banking. World Scientific, Singapore, pp 261–321

Part III
Ensuring Safety by

Chapter 10
Testing the Donor

Alan Kitchen and John Barbara

Introduction

The transplantation of tissues from donor to recipient is a fundamental part of medicine today. A range of tissues including bone, skin, tendons, heart valves, corneas etc. are collected from suitable donors and transplanted into those patients for whom tissue transplant would have clinical benefit.

In the same manner as blood transfusion, transplantation is not without risks, one important risk being that of the transmission of infectious disease via the transplant; transplantation transmitted infections (TTI). Although it is clear that not all tissues carry the same risk, such distinctions cannot be easily quantified and all tissues should be treated in the same way in terms of the infectious disease screening applied.

Thus, in the same way as for blood donations, all tissue donations must be screened for infectious diseases prior to release for clinical use. Although the basic principles applied to the screening of tissue donations are the same as for blood donations, or any other type of donation of biological products, there are some important differences between blood and tissue donations which result in differences in the actual screening strategies applied. Firstly, and most importantly the source of tissue donations includes both living and deceased (cadaveric) donors, and secondly, whilst the specific infectious agents screened for are broadly the same, the specific markers of infection used for the individual infectious agents may differ.

A fundamental issue when considering the screening of tissue donations for transmissible infectious agents (TIA) is the status of the donor, living or deceased. The screening of tissues collected from living donors for TIAs is effectively no different to the screening of blood donations, in that the blood samples used for screening are collected by normal venepuncture from living individuals. The

A. Kitchen (✉)
National Transfusion Microbiology Reference Laboratory, NHS Blood and Transplant,
Colindale Avenue, London NW9 5BG, UK
e-mail: alan.kitchen@nhsbt.nhs.uk

G. Galea (ed.), *Essentials of Tissue Banking*, DOI 10.1007/978-90-481-9142-0_10,

characteristics of the samples are identical and the screening is performed using assays specifically designed, developed and supplied to test blood samples collected from living individuals. It is the screening of tissues from deceased donors, where the blood samples are collected post-mortem (non heart beating), that presents some specific problems that need to be overcome to ensure accurate and reliable screening results. These problems all surround the overall quality of the samples obtained and their validity as suitable substrates for the assays used; the time post-mortem of collection, the integrity of the sample, volume, level of haemolysis, biochemical changes post-mortem etc. However, as long as these problems are acknowledged and dealt with, and resolved in the right ways, the overall quality of the screening programme should be the same whether the samples collected to screen the tissue donations are from living or deceased donors.

Similarly, the screening of the donors themselves, the donor selection process, is also determined by the nature of the donor. The selection process applied to living donors is similar to that applied to blood donors, except that such donors do not generally volunteer in the same way as an "altruistic" blood donor. Living tissue donors may often have been approached directly and a direct history may therefore be taken from them. However in the case of deceased donors, although the process is the same, the history is second hand and may not be totally correct or accurate [1]. Thus the inherent risk associated with deceased tissue donations must necessarily be higher than that of living donations and the screening strategy developed accordingly.

This text looks at the screening of tissue donations for the range of infectious agents that may be present in the donor at the time of collection of the tissue(s) and which are therefore likely to be present in the donations collected and consequently likely to be transmitted through the transplantation of the donated tissues. These agents include viruses, bacteria, parasites and prions. The subject of the bacterial/fungal contamination of tissue products as a result of the collection, processing and storage processes is not covered in this particular text.

Range of Infectious Agents

The infectious agents that tissue donations need to be screened for are essentially the same as for blood donations, as are the specific markers used as screening targets for the individual infectious agents. However it may be that for some infectious agents additional markers may be included to increase the sensitivity of screening.

Table 10.1 lists the range of infectious agents that have been reported to be transmitted via blood and blood products. Although only a few of these agents have been reported to be transmitted via tissue transplantation, it is reasonable to consider that all of these agents could and would be transmitted if present in a tissue donor and in the donated tissues. Thus the same approaches in terms of identifying specific donor risk need to be followed, albeit with the obvious issues in the pre-selection of deceased donors.

Table 10.1 Infectious agents currently reported to have been transmitted via blood transfusion or tissue transplantation

Viruses

Hepatitis viruses
Hepatitis A virus
Hepatitis B virus
Hepatitis C virus
Hepatitis D virus (requires co-infection with HBV)
Hepatitis E virus

Retroviruses
Human immunodeficiency virus 1 and 2 (plus other subtypes)
Human T-cell leukaemia virus I and II

Herpes viruses
Human cytomegalovirus
Epstein–Barr virus
Human herpesvirus 8

Parvoviruses
Parvovirus B19

Miscellaneous viruses
GBV-C: previously referred to as hepatitis G virus
TTV
West Nile virus
Dengue virus
Rabies via organs

Bacteria

Endogenous
Treponema pallidum (syphilis)
Borrelia burgdorferi (Lyme disease)
Brucella melitensis (Brucellosis)
Yersinia enterocolitica/Salmonella spp.

Exogenous
Environmental species: staphyloccocal pp./pseudomonads/*Serratia* spp.
Rickettsiae: *Rickettsia rickettsii* (Rocky Mountain spotted fever), *Coxiella burnetii* (Q fever)

Protozoa

Plasmodium spp. (malaria)
Trypanosoma cruzi (Chagas' disease)
Toxoplasma gondii (toxoplasmosis)
Babesia microti/divergens (babesiosis)
Leishmania spp. (leishmaniasis)

Prions

variant Creutzfeld Jakob Disease (vCJD)
[classical CJD has been transmitted by corneas and dura]

The actual transmissible infectious agents that are of specific interest can be considered in three categories:

- those for which all tissue donations must be screened – mandatory
- those which are recognised and commonly encountered threats, but for which only specific donations need to be screened because of specific identifiable risk – discretionary (UK terminology)
- those which are rare or unknown threats and which are dealt with on a case-by-case basis – others

In addition to defining the infectious agents themselves, the specific target marker of infection for each infectious agent must also be defined. The available targets vary according to the infectious agent and theoretically there are always 3 targets for any infectious agent. However it is the suitability of each of these targets in the context of the screening of what should be low risk donors, that has to be considered, from the perspective of the biology of the agent and the host response to infection, and the sensitivity and predictiveness of the available tests. Figures 10.1, 10.2, 10.3, 10.4, 10.5 and 10.6 show the different plasma markers that can be detected following infection with HBV, HIV and HCV respectively. The figures depict the relative

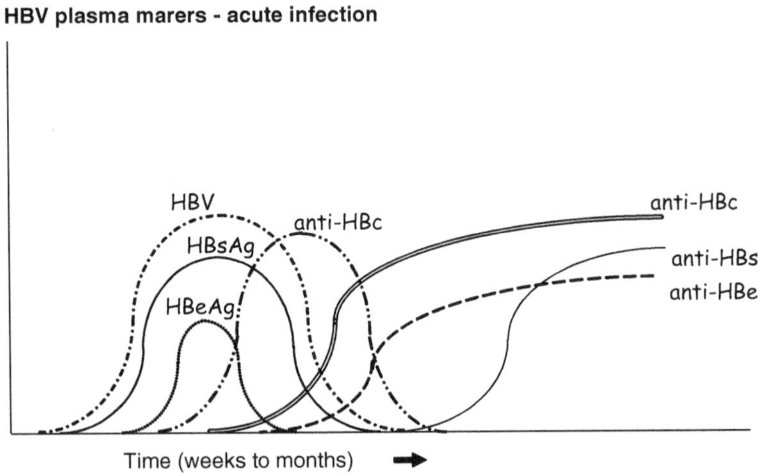

Fig. 10.1 HBV plasma markers – acute infection

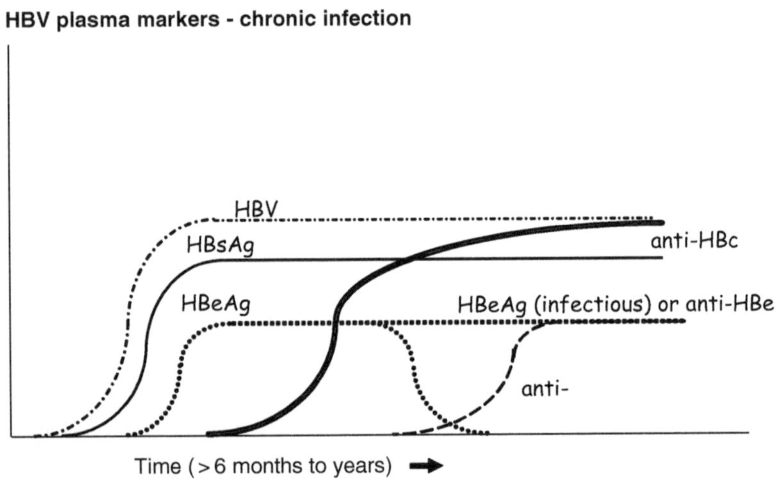

Fig. 10.2 HBV plasma markers – chronic infection

timeframes for the appearance of the different markers for each of these agents, and from these the most useful markers in terms of reliable detection of infection can be determined. A pre-requisite for any effective screening programme is always a thorough understanding of the biology of the agent and course of the infection in the host.

Even though it could be argued that some tissue products are processed in ways that remove or inactivate any infectious agent that could theoretically be present, the use of tissue products collected from a donor who may be infected is not considered to be appropriate practice, irrespective of the processing methods used. No pathogen inactivation or removal methods can be considered to be totally effective

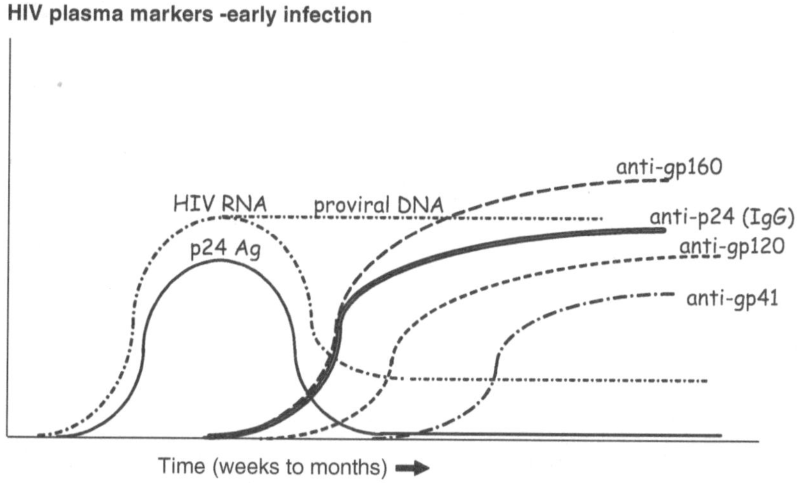

Fig. 10.3 HIV plasma markers – early infection

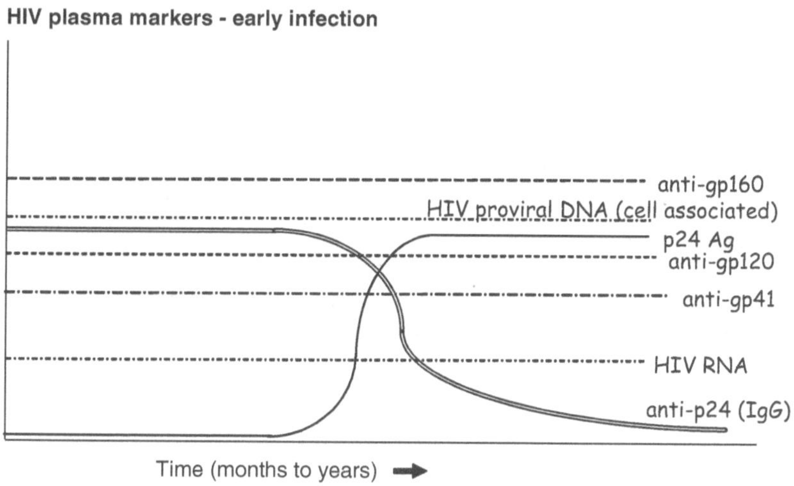

Fig. 10.4 HIV plasma markers – late infection

and therefore must be used in conjunction, not instead of, screening. If a donor is known to be infected with any of the transmissible infectious agents there is a clear risk of transmission and therefore the products should not be released for clinical use.

Mandatory Screening

The infectious agents that are currently considered to be mandatory for the screening of tissue (and blood) donations in most countries with developed healthcare systems

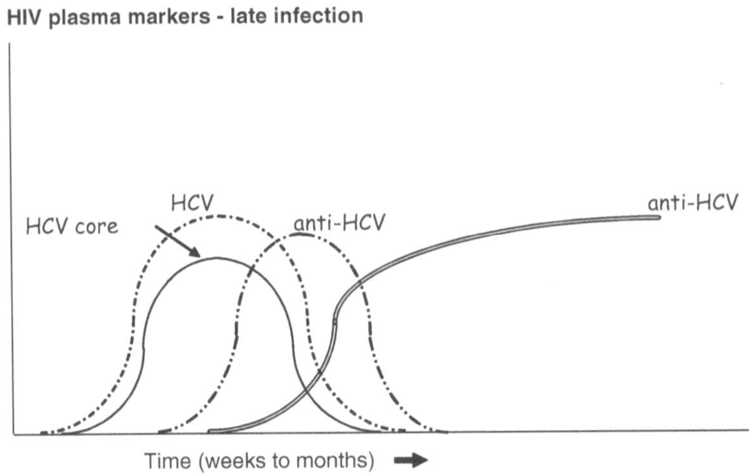

Fig. 10.5 HCV plasma markers – acute infection

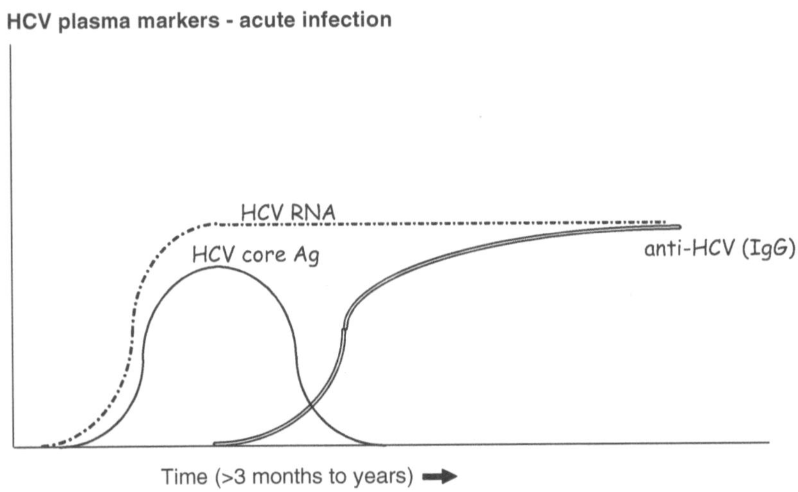

Fig. 10.6 HCV plasma markers – chronic infection

and formal regulatory authorities are hepatitis B virus (HBV), hepatitis C virus (HCV), human immunodeficiency viruses 1+2 (HIV), human T-cell lymphotropic viruses I+II (HTLV) and *Treponema pallidum* (causative agent of syphilis). These 5 infectious agents are those for which there is published, irrefutable evidence of transmission through blood and other body fluids, and therefore the potential for transmission through residual blood in tissue products. The agents are generally present in populations worldwide, although the prevalence and incidence does vary significantly from country to country, and present such a general risk that, in theory, any donor could be infected, irrespective of absence of (known) risk or symptoms.

It is important to understand that risk is something that the pre-donation questioning and selection process seeks to identify, but donors may have been exposed to an infectious agent through a route that they were not aware of. In addition, all of these infectious agents do produce asymptomatic sub-clinical and chronic infections, and donors may appear to be fit and well at the time of donation whilst being infected and thus infectious. This is especially important in the screening of cadaveric donors as they obviously cannot be questioned directly, and the selection process is dependent upon what family and close contacts believe, which may not necessarily be the truth.

All donations are therefore theoretically at risk of being infected with one or more of these agents and so blanket screening of all donations is necessary. However, as mentioned above, as well as identifying the infectious agents, the specific screening target(s) for each infectious agent must be identified. Table 10.2 lists these specific screening targets and the mandatory screening requirement for all donations collected by the UK blood services.

Discretionary Screening

Those infectious agents for which screening can be considered to be discretionary, i.e. not all donations require screening, comprise agents for which again there is published, irrefutable evidence of transmission through blood and other body fluids and which can cause significant disease, but where risk is, and can be more easily, mitigated because the agent is not endemic or naturally present in the general population in that country, and where there are particular risk factors for infection that can be specifically identified in a donor. Specific examples of infectious agents that can be defined as discretionary are West Nile Virus (WNV), *Plasmodium* spp. (causative agents of malaria), *Trypanosoma cruzi* (causative agent of Chagas' disease). To be considered as a potential risk there is an absolute requirement for exposure of the donor to the virus (WNV) or parasite (*Plasmodium/ Trypanosoma*) by either living in, or visiting, an endemic area. Donors who are at risk can therefore be easily and specifically identified at the point of donation and only donations from such donors require this additional screening.

Other Screening

In addition to the well known, characterised and commonly screened for infectious agents, there is always the potential for the transmission of other infectious agents.

- These may be existing agents which have a generally low prevalence and incidence, often tightly geographically restricted, are rarely transmitted, but which nonetheless may be transmitted if present in the donor at the time of donation.

Table 10.2 Current mandatory screening of donations within the UK blood services

Marker/donation	HBs Ag	HIV Ag/Ab	HCV Ab	Syphilis Ab	HTLV Ab (S)	HTLV Ab (P)	HBc Ab	HCV RNA (P)	HCV RNA (S)	HIV RNA (P)	HIV RNA (S)	HBV DNA (S)	HCMV Ab	HCMV DNA
Blood	X	X	X	X		X		X						
Surgical tissue (one sample)	X	X	X	X		X	X	X		X		X		
Surgical tissue (two	X	X	X	X		X	X							
samples, 180 days apart)	X	X	X	X		X	X							
Deceased donor (adult)	X	X	X	X	X		X		X		X	X		
Deceased (I)	X	X	X	X	X		X		X		X	X		
(neonate) (M)	X	X	X	X		X	X	X		X		X		
Stem Cells	X	X	X	X	(X)[a]	(X)[a]	X							
Allogeneic cord blood	X	X	X	X	X		X		X		X	(X)[b] HBcAb+	X	(X)[c] CMV Ab+

[a]Singleton HTLV is needed in "high risk" HTLV areas.
[b]Required if the donation is found to be HBcAb+, irrespective of HBsAb status.
[c]Required if mother is CMV Ab+.

- There may be emerging agents that are transmissible, have been rarely transmitted in the past but where their incidence and prevalence are starting to increase, either through increasing incidence in endemic countries, through spread into previously non-endemic countries or through increased travel into endemic countries.
- There may be yet unidentified infectious agents that would be transmissible but which at this time have not appeared as the causes of any identifiable transfusion or transplantation associated infections.

Table 10.1 can be considered as a baseline list of infectious agents that are known to present a threat because of reported blood transfusion transmissions. Although the agents listed have been transmitted through blood, there may have been several transmissions or only a handful of cases globally especially the more "exotic" agents listed. However, to provide a definitive list of all possible current and future infectious threats is clearly impossible. In terms of potential candidate agents, meeting the following criteria would identify those particular agents that could be transmitted through tissue transplantation, if present in the donor at the time of donation. To be considered a viable threat the agent must be:

- Present in the donated tissue, in an infectious form
- Stable in the donated tissue/processed processed under the storage conditions normally required for such products
- Circulating in the donor at the point of collection of the tissue, yet without giving rise to any identifiable symptoms in the donor

This does narrow the field somewhat, and it must be made clear that not every infectious agent that afflicts man is a potential transfusion/transplantation threat. Indeed the number of potential threats is only a small proportion of those infectious agents currently known to man. Nonetheless, there are a few well characterised agents that, whilst not a major concern at present, may spread into currently non-endemic areas and therefore need to be monitored. These include viruses such as SARS, Chikungunya virus, Lymphocytic Choriomeningitis Virus (LCMV), and parasites such as *Leishmania* spp.

One major problem that may influence any screening programme for an infectious agent are the potential limitations either in the appropriate screening target(s) available and/or the corresponding screening tests being available. This is may significantly hinder impact on the ability to develop effective and appropriate screening strategies for some agents, resulting in a greater reliance on donor exclusion rather than in-vitro screening (something that is not always possible or effective with cadaveric donors/donations).

Severe Acute Respiratory Syndrome (SARS)

Severe acute respiratory syndrome (SARS) is a respiratory infection caused by a recently emergent coronavirus (SARS-CoV). The disease has severe morbidity and mortality, but presents with non-specific signs and symptoms and there is no clear-cut diagnostic approach to prospectively identifying cases prior to the appearance of symptoms. The virus has a viraemic phase of 4–8 days, which precedes symptoms and then persists into the symptomatic phase. Viraemic individuals may transmit if blood or tissues are collected during the early asymptomatic phase of infection. Screening of individuals for SARS-CoV is possible using molecular techniques but identification and deferral of donors who may have been potentially exposed to the virus is currently the most effective way of minimizing risk of transmission.

Chikungunya

Chikungunya is an arbovirus (viruses transmitted by insects) that is transmitted primarily by mosquitoes, usually of the *Aedes* spp.. The illness is an acute symptomatic disease that characteristically begins with rapid onset of joint pains and may or may not be accompanied by muscle pain, high fever, conjunctivitis, and a rash. The severity of the illness varies and tends to be less severe in children. Most infected individual recover after a few weeks but a small proportion may have residual chronic joint pain for some years. Infection is not usually fatal, but may be a contributing factor in the death of some patients with pre-existing underlying disease. The disease occurs throughout Africa, Asia, parts of southern Europe, anywhere where the carrier mosquito species are found. More recently there have been a number of outbreaks on the islands in the Indian Ocean, one in particular on Reunion resulting in the cessation of blood collection until the outbreak had passed.

The virus has a viraemic phase prior to and during the symptomatic phase. Viraemic individuals may transmit if blood or tissues are collected during the early pre-symptomatic phase of infection. Screening of individuals for Chikungunya is possible using molecular techniques but identification and deferral of donors who may have been potentially exposed to the virus is currently the most effective way of minimizing risk of transmission. However the overall risk of the transmission of Chikungunya through tissues is very low as the disease is of acute onset and generally highly symptomatic. In 2006, 133 cases of imported Chikungunya were recorded in the UK. All of these had a clear history of risk and most had clinical symptoms.

Lymphochoriomeningitis Virus (LCMV)

Lymphocytic choriomeningitis virus (LCMV) is a virus carried by rodents that causes lymphocytic choriomeningitis (LCM). It presents as an aseptic meningitis, encephalitis or meningoencephalitis, although infection is often asymptomatic or with only mild febrile illness. The disease is rarely serious although infection during pregnancy can lead to severe disease in the infant. Infection is most commonly acquired through contact with infected rodents, often household pets, their urine, droppings bedding etc.. Transmission can also occur via bites and through open wounds where there has been contact with contaminated material. Human-to-human transmission has not been reported except for vertical transmission from mother to foetus. However LCMV has been identified as being transmitted through organ transplantation [2], although cases of acute LCM in such donors would be very rare events. Screening for LCMV infection is possible, but molecular screening would be needed to identify viraemic donors.

Leishmania

Leishmaniasis is a parasitic disease that is found in parts of the tropics, subtropics, and southern Europe. It is caused by infection with *Leishmania* parasites, which are spread by the bite of infected sand flies. There are several different forms of leishmaniasis in humans, the most common forms being cutaneous which causes skin sores, and visceral (Kala Azar) which affects internal organs (e.g. spleen, liver, and bone marrow). The factors determining the form of disease include leishmanial species, geographic location, and immune response of the host. Although generally symptomatic, asymptomatic infections do occur and it is possible for parasitaemic individuals to present as blood or tissue donors, although there are no reported cases of transmission through blood or tissues. Screening for leishmaniasis is not straightforward as antibody responses may be low or absent in cutaneous cases, although screening is possible using molecular techniques. However identification and deferral of donors who may have been potentially exposed to the virus is currently the most effective way of minimizing risk of transmission

Screening Strategy

To be effective, in-vitro screening requires an effective screening strategy to be developed to ensure that the testing performed is targeted correctly and the outcomes are used effectively and appropriately.

Probably the first issue to be considered in the development of strategies for the screening of tissue donations is the fact that tissue donations can be divided into those from living donors (surgical tissues) and those from deceased donors (deceased tissues). The fundamental differences in the origins of the tissues, and thus the approaches available to "screen" the donors, means that the two have to be considered separately, and separate strategies developed. Whilst the core screening needs are essentially the same, the obvious restrictions in the donor selection process that can be applied to deceased donors means that there is a much greater reliance on in-vitro laboratory testing to determine the safety of the tissues collected. In the case of surgical tissues, however, the donor can be treated in virtually the same way as any blood donor, with a formal donor selection process prior to donation to identify any specific risks and determine the overall suitability of the donor as a source of a tissue donation. Although laboratory testing is still the final decider in the release of the donation for clinical use, experience from the selection of blood donors has shown the value of the donor selection process in reducing the risk of collecting a donation from a "high risk" donor, namely a donor who may have been recently infected, and is infectious, but where the infection may not be detectable at that time with the tests in use – the window period.

There are two basic elements to a screening strategy, firstly defining the infectious agents and the individual specific screening target(s) for each infectious agent, and secondly the actual testing algorithm adopted.

As discussed above, the range of infectious agents that need to be considered to ensure tissue safety is similar to those for blood and generally well defined. Likewise the specific screening targets for each of these infectious agents are also well defined. However, these targets may differ depending upon the relative inherent risk associated with different donation types, in the case of tissues – deceased or surgical. Additional screening is added where the inherent risk is higher. Table 10.2 outlines the mandatory screening requirements for blood, tissue and stem cell donations collected and processed within the UK transfusion services. Although there is a common core, there are quite different additional requirements for different donation types. Defining the specific screening targets is therefore a fundamental part of any screening strategy, and has to reflect the specific infection risks in the donor population from which the tissues are collected.

The second key element of the screening strategy is the actual screening algorithm adopted. The algorithm essentially defines how the screening assays are used in terms of the initial and (any) repeat testing, and how the final screening result is determined. Again, this area is very similar to that for blood screening. The algorithms that are used for blood screening apply equally to tissue, and any other donation, screening. However it is essential to define the most appropriate algorithm for the particular screening performed, and it is here that blood and tissue screening may diverge. There are really only three possible algorithms that can be used for the screening of donations from low-risk donor populations to determine their fate, i.e. suitable for use or discard:

Algorithm 1 Initial screening test (Assay A) and use the results obtained to determine the donation fate. Negative – suitable for clinical use, Reactive – discard the donation

Algorithm 2 Initial screening test (Assay A) and repeat any initial reactives, in duplicate, using the same test. Use the 2 out of 3 rule (the 2 identical results out of the 3 results obtained are considered to be the true screen result) to determine the donation fate. Negative – suitable for clinical use, Reactive – discard

Algorithm 3 Initial screening test (Assay A) and repeat any initial reactives, in duplicate, using a different screening assay (Assay B). Use the 2 out of 3 rule to determine the donation fate. Negative – suitable for clinical use, Reactive – discard

The 3 algorithms are clearly quite different and have different uses, depending upon the type of donations being screened, the levels of infection in the donor population, and the complexity and effectiveness of the quality management system. Although all 3 algorithms are effective and will ensure similar levels of product safety, it is the associated specificity that is perhaps of more interest and relevance when dealing with tissue donations. Importantly all 3 screening algorithms have the same sensitivity as all take the same approach in respect of the initial screen, and the acceptance of screen negative donations as being suitable for release for clinical use. Differences in sensitivity are solely due to the sensitivity of the screening assays used.

In the case of blood donation Algorithm 2 is the one used by the majority of transfusion services in countries with developed healthcare systems, and offers an effective approach, with an acceptable level of specificity. Although every donation is valuable, it is appropriate to point out that blood donations are more numerous than tissue donations, donors can donate regularly and it is far easier to replenish stocks. Thus the specificity of the screening and the algorithms, whilst important, is not so critical for blood donations, and the algorithm used reflects this. Indeed in those countries that use Algorithm 1, the unnecessary wastage of blood donations is even higher. However, in the case of tissue donations, and especially deceased donations where a single cadaver could provide a significant number of different tissues, the specificity of the screening algorithm is of far greater significance as repeatable screen reactivity would result in the discard of the products, even if that reactivity could be subsequently proven, through confirmatory testing, to be non-specific. Algorithm 3 is based upon the use of a second assay, Assay B that has at least equal sensitivity to Assay A, that is used to test the initial screen reactives. The algorithm is based on the fact that most non-specific reactivity is assay specific, and thus the majority of non-specific reactivity seen with Assay A, are unlikely to be seen with Assay B, whilst any true reactivity will always be seen in both assays. The algorithm increases the specificity of screening significantly, reducing the unnecessary loss of donations just to those that are reactive with both assays, generally those where infection is more likely to be confirmed. However Algorithm 3 is more challenging, both technically and from a quality management perspective, and is not an algorithm that is felt to be advantageous in a mass screening environment, screening blood donations. The major issue here is that a lot of tissue donations are collected, tested and processed by transfusion services and the screening approach used is that which was developed for blood donations, and may not be the optimal for other donation types, especially deceased donations. Thus consideration must be given to implementing the most appropriate algorithm for the product type, rather than the blanket use of algorithms that are primarily designed for blood donations, and where a lower level of specificity may be more acceptable.

This is particularly critical for deceased tissue donations where the samples are collected post-mortem and which are not the optimum substrate for the screening assays used and where overall the specificity of screening may be lower than that of blood donation screening. Although a number of authors have reported problems with the specificity of deceased screening programmes [3–7], the deceased tissue screening programme currently in place within the English National Blood Service is not encountering these levels of screen reactivity and the overall specificity of the screening programme is relatively high.

Sample Quality

The quality of the blood samples used to screen the tissue donations is clearly important. Poor quality samples produce poor quality results, with the potential resultant loss of sensitivity and/or specificity. The first issue to consider is the definition of quality as applied to blood samples collected for the purpose of screening tissue donations for release for clinical use, and here the difference between deceased and surgical tissues is very important.

Sample quality can be considered to include all of those attributes that may influence the overall reliability and accuracy of the screening results obtained. Importantly these include the condition of the sample including its age, the constituents of the sample (representativeness), and the volume. There are other, important, issues such as donor and sample ID etc., but these are outside the scope of this particular chapter.

Samples from Living Tissue Donors

The difference between deceased and surgical tissue donations is important, in relation to sample quality and in-vitro screening, as the blood samples collected from surgical donors are collected from living ("heart-beating") individuals and are thus the normal sample type that the screening assays are designed to be used for, and also should always be of sufficient volume for the screening required. Therefore the screening of surgical donations, notwithstanding the previous discussion about specificity, can be and are performed using the same systems and approaches as those used for blood donations. There should not be any differences in the nature of the blood samples collected or the reliability and accuracy of the results obtained.

Samples from Cadaveric (Non Heart-Beating) Donors

Unlike samples from living donors, blood samples from deceased tissue donors, unless the sample was collected whilst the donor was still heart-beating, cannot be expected to necessarily behave in the same way as those from heart-beating individuals, and therefore the screening performed and the results obtained have to be considered more carefully, and their validity established. A key issue, however, is not so much the fact that the blood sample is from a cadaver, but rather the representativeness of that sample, its suitability for use with the screening assays in use and therefore its ability to fully reflect the infectious disease status of the donor.

It is the nature of the blood sample collected from a cadaver that presents the major challenges. Blood samples collected post-mortem may be different from those collected ante-mortem for a number of reasons. The sample is collected from a now static fluid that is settling into its cellular and liquid phases, in which coagulation may have occurred with the resultant biochemical changes including red cell damage, and where cell death is taking place with the release of a wide range of cell chemicals into the surrounding tissue fluids. The overall results of this may be gross haemolysis and/or other changes to the "blood" collected, including degradation of the screening targets that would be present in blood of an infected individual, and/or the release of substances that may be inhibitory to the screening assays used. However the changes that occur post-mortem obviously vary between cadavers, depending upon factors such as: age, cause of death, time between death and the cold-storage of the cadaver, time pre-mortem under direct clinical care and the interventions used, other underlying conditions (that do not de-bar tissue donation), and time post-mortem of sample collection. Thus the occurrence of post-mortem changes per se do not necessarily make any sample collected a less suitable substrate for screening. The screening of post-mortem samples from deceased tissue donors has been performed for many years in many countries with few, if any, known adverse outcomes. In addition assays can be validated for use with deceased samples, and indeed a few commercially available infectious disease screening assays for blood-borne viruses have undergone some limited validation work by the manufacturer with the resultant inclusion of claims that the assays can be used to screen deceased samples collected under certain conditions.

An important factor in the screening of deceased samples is the time, post-mortem, that the sample was collected. Clearly the sooner after death the sample is taken, the less likely it is to have been effected by any post-mortem changes. Interestingly however, currently there are few published papers that look critically at the time post-mortem for sample collection. Regulatory requirements within the EU require sample collection within 24 hours post-mortem, and in the US there is an expectation that tissues would be collected as soon as possible post-mortem, and within 24 hours, but this is currently not subject to any specific regulation. There are published studies on aspects of deceased testing that include samples collected up to 48 hours post-mortem and without any problems being encountered [4, 8], but specific studies on sample suitability related to time of collection are lacking.

Additionally there is the problem of possible haemodilution [9]. Very often the deceased donors were on life-support systems prior to their death and this may have involved either volume replacement with a range of inert volume expanders or by transfusion, in both cases potentially diluting out the constituents of the patients' own blood and therefore reducing the representativeness of the sample in terms of reflecting the true status of the donor. The immediate pre-mortem history of the donor must be known if there is any likelihood of adverse effects due to haemodilution, and at the very least the sample collection site chosen to try to minimise the risk of collection of a sample containing a significant amount of resuscitation fluid. There is a reported case of the failure of screening to identify an HIV infected tissue donor specifically because of massive fluid infusion shortly before death with the resultant dilution of HIV antibodies to below that detectable. Interestingly the antibody level returned to that normally seen, and detectable, 48 hours later as the tissue fluids naturally re-distributed themselves [10]. However, most of the studies performed [9] have indicated that the modern screening assays used will tolerate a significant dilution of plasma before approaching the point at which a false negative result may occur. This situation is most likely to occur following significant acute blood loss and this fact will be apparent when assessing any potential deceased donor. There are FDA guidelines which seek to help clarify the situation, and these require tissue banks to develop algorithms to determine the degree of haemodilution and the action to be taken in terms of the validity of screening results [11] Within NHSBT there is a clear algorithm that can be followed in cases where haemodilution has occurred, that enables the degree of haemodilution, and thus sample suitability, to be assessed.

The actual physical process of obtaining a sample from a cadaver is very different and sample volume is often a major challenge in performing the screening required. As described above the blood settles out across the cadaver, both from the periphery into the larger vessels, and the cells settle out below the fluid component. Suitable sites for sample collection must therefore be identified, suitability depending on a number of factors. These include the presence of a suitable volume of fluid in the vessel and the accessibility of the vessel to facilitate sample collection, the probability of any resulting damage to tissues of interest, for example samples collected through cardiac stab, and the collection process itself must ensure that a suitable sample is obtained, i.e. sufficient serum/plasma to be able to perform all of the screening tests required. In addition the possibility of bacterial contamination of both the sample and the tissue itself, either due to bacteria already present in the cadaver or through the collection process itself, must always be considered. However grossly bacterially contaminated samples are usually very obvious and the suitability of the sample questioned, although bacterial contamination of the tissue itself, prior to retrieval, would not be so easily identified.

There is also a concern that if for example the sample is obtained through a cardiac stab one may damage the valve/introduce infection at the site.

Thus although a post-mortem sample from a deceased tissue donor may not be the optimum substrate for the screening assays used, if collected carefully and under controlled conditions the sample collected may resemble that obtained from any heart-beating individual with few, if any, post-mortem changes that would have any adverse effect on any screening performed. Certainly the majority of approximately 2000 samples/yr referred to the author's laboratory, the single national laboratory responsible for the screening of all deceased tissue donations collected within NHS Blood and Transplant (NHSBT) and which also provides a screening service for some non-NHSBT Tissue Banks, are suitable for screening and do not give rise to any problems beyond a slightly lower overall specificity [12].

Thus, whilst samples collected post-mortem from deceased tissue donors can be suitable substrates for many screening assays, the collection conditions must be properly controlled to ensure that they are as representative of a pre-mortem sample and the serological status of the donor as possible. In addition, the screening systems developed must be specifically designed to generate as reliable and accurate results as possible from this group of samples.

Screening

The in-vitro laboratory screening performed can be divided into serological and molecular. To be able to understand the specific role and value of the two different approaches, the basic biology of infection must be understood. Following exposure to an infectious agent there is a period during which the infectious agent either enters the body's cells and starts to replicate or is captured and eliminated by the body's passive defence mechanisms. Once the agent has entered the body's cells and started to replicate it is highly likely that a productive infection will follow. There is a period during which the infectious agent continues to replicate locally until there is sufficient to start to spread to other suitable target sites within the body. During this time levels are low and generally not detectable. However as the agent starts to spread (generally via the circulation) when levels are high enough it can then be detected. At this point nucleic acid can start to be detected, and depending upon the infectious agent, antigen may also be detectable. Subsequently as the agent spreads the immune response is initiated, symptoms appear and antibody levels start to rise and become detectable. However, again depending upon the agent, at this point nucleic acid levels in the circulation may start to fall as the immune system starts to combat the infection and a balance ensues. In the case of the majority of the blood borne infectious agents the infectious agent then sequesters itself in its preferred cells in the body and a chronic or long-term infection develops, the antibody produced simply indicating infection at some time and possibly mediating the infection, but not necessarily conferring any immunity to the agent. In the case of other infectious agents, those that give rise to acute infections only, the appearance of antibody in the circulation marks the start of the resolution of infection and the clearance of the agent from the body, and subsequent immunity, for at least a short period, to that particular infectious agent.

Serological screening can be described as the conventional approach, looking for the presence of specific serological makers of infection in blood samples collected from the donors. This approach has been in use for many years and offers sensitive, reliable, fully process-controlled and cost-effective screening which will identify the vast majority of donors who are genuinely infected (early, recent, ongoing and old infections). Serological screening is relatively cheap and although it requires dedicated, often very specific equipment, this equipment is not particularly "high-tech", nor particularly expensive when considering the work rates achieved and its overall reliability. Molecular screening is more recent in its origin and has only relatively recently been applied to mass donation screening programmes. Indeed the mass screening applications are still being developed, and currently the bespoke screening systems are primarily designed for blood rather than other donation types. Molecular screening targets the genomic material of the infectious agents and works through the cyclic amplification of low levels of target (normally undetectable) to high levels which can then be more easily detected. It offers a (theoretically) higher sensitivity than serological screening for identification of early infections, but at the same time is less effective in detecting existing/old infections. Thus molecular screening should only be considered as an addition to, not a replacement for, serological screening. In addition it is relatively expensive when compared to serology, as it is a more "high-tech" approach utilising new and developing technologies and requiring dedicated specialist equipment for the automated mass screening of donations now required by many countries. A major question that is still to be addressed is whether cost of molecular screening is justified by the incremental benefit of molecular screening over serology alone.

Serological Screening

Serological screening is that performed to look for the presence of specific serological markers of an infectious agent that are found in the blood as a result of the infection. These markers are either specific antigens produced by the infectious agent and which generally appears on the surface of the agent, but in some cases may be expressed on infected cells, and/or specific antibodies produced by the individual's immune system in response to the infection. Table 10.2 outlines the serological markers for which tissue donations collected within the UK transfusion services are screened.

Although the immune response to any infectious agent is often associated with protective immunity, indeed the whole purpose of vaccination is to stimulate the body's immune response to produce protective antibodies, in respect of the majority of the blood borne infectious agents that are transmissible through blood and tissues, except in the case of HBsAb which is a protective neutralising antibody, the presence of antibody is not indicative of protection, rather it indicates infection and the presence of the infectious agent in that individual. A common and important feature of most of the relevant blood-borne infectious agents is that they are all persistent infections which generally do not resolve naturally. Although approximately 40% of cases of HCV do resolve, such individuals are serologically indistinguishable from persistent infections, and thus the presence of HCV antibody cannot differentiate between infectivity versus immunity.

There are a number of different types of immunoassay available, but all have broadly the same principles. However not all are suitable for screening cadaveric samples, often due to technical (sample and sampling) rather than scientific issues. In general immunoassays comprise a solid phase and a liquid phase, and the assay is performed in a number of stages, each of which has an incubation period and a wash to remove the excess materials at the end of the incubation. The solid phase carries the immobilised components of the assay, those designed to capture the specific target that may or may not be present in the liquid phase (sample). In the microwell format assays, the well is coated with either antigen (Ag) or antibody (Ab), depending on the specific target (Ag to detect Ab, Ab to detect Ag). Captured target is then detected using a conjugate which detects any Ag/Ab complex formed and has an enzyme attached to it. This enzyme triggers a colour reagent that then signals the presence of the conjugate and thus the target through the development of colour which can then be measured.

Most of the assays used today to detect both antigen and antibody, either singly or in combination, depending on the particular infectious agent, are highly sensitive and specific immunoassays that are designed to detect specific target as early as possible in the infection process, and run on automated platforms with high levels of process control. This ensures highly accurate and reproducible results and thus the overall reliability of the screening process. However, at the same time this aspect is also a problem, certainly in the case of deceased samples, as the automated screening platforms are designed specifically for the screening of blood donations. This is primarily to do with the automated sample detection systems rather than the assays themselves, with for example, haemolysed, low volume, or slightly diluted samples all being rejected as unsuitable for processing. As identified previously, it is here that surgical tissue donations align virtually completely with blood donations as the samples for both are collected from living individuals and are thus generally good quality samples, the optimum for these systems. Whereas deceased samples are commonly rejected by such systems as they fail to be detected properly by the system and therefore not sampled and not tested. At this time there is little option but to use separate screening strategies for deceased samples, using assays and systems that can be used and have been specifically validated for the screening of deceased samples.

There are two potential performance issues that need to be considered for any screening programme, but which may be of more relevance to serological screening, and specifically to the serological screening of deceased donations; sensitivity and specificity. A past major reported issue has been the specificity of the serological screening of deceased donations. Many authors report high reactive rates with the serological screening assays in use and the consequent high losses of donations. Whilst it is true that some deceased samples do present challenges when performing serological screening, mainly those that are visibly of poor quality, our experience is that the specificity of the screening assay is not the problem that some report, rather our concern is over the overall sensitivity of the screening programme. It is important that the assays used are validated specifically for use with deceased samples to determine both any specificity issues and any sensitivity issues. The problem is that specificity issues can be seen as high screen reactive rates which are not subsequently confirmed, wasteful but not affecting safety. However any sensitivity issues would only be seen once an infected donation had been transplanted and the recipient infected. Thus an important part of assay evaluation and validation for the screening of deceased blood samples is to demonstrate that the deceased nature of the sample does not reduce the sensitivity of the assay. Unfortunately this in itself presents problems as suitable cadaveric samples are simply not available and indirect methods have to be used [13].

Molecular Screening

Molecular screening is that performed to detect the presence, in the blood, of the nucleic acids of transmissible infectious agents. The presence of viral nucleic acid specifically indicates the presence of virus, and, in most cases, infectivity. At this time the infectious agents that are targets for molecular screening are hepatitis B virus (HBV DNA), hepatitis C virus (HCV RNA) and human immunodeficiency virus (HIV RNA), and, for one particular product human cytomegalovirus (HCMV DNA), Table 10.2 outlines the viral nucleic acids for which tissue donations collected within the UK transfusion services are screened. Essentially the molecular screening used to detect low levels of virus can be broken down into two elements, the extraction and purification of any nucleic acid in the sample and then the subsequent specific amplification and detection of any target nucleic acid present. Although viral nucleic acid may be present in the sample from the donor, and even if there are thousands of genome copies present, the overall amount of "native" nucleic acid present is still far too small to be able to detect directly with the technology currently available. Instead, the approach taken is to isolate the nucleic acid present in the sample and then amplify any specific target that may be present to a level that can then be detected either by direct visualisation using nucleic acid binding fluorescent dyes or through enzyme action with the generation of a chemiluminescent signal, both in proportion to the amount of nucleic acid present. There are currently 2 main molecular methodologies used in the molecular screening of donations for infectious agents, the polymerase chain reaction (PCR) and transcription mediated amplification (TMA). Although this chapter is not the appropriate place to provide an in-depth review of these methodologies, an overview follows.

PCR is a temperature dependent method that allows exponential amplification of short DNA sequences present within a longer double-stranded DNA molecule, in this case the genome of the infectious agents being screened for. It entails the use of a pair of primers, short sequences of nucleotides that are complementary to a defined sequence on each of the two strands of the DNA. These primers are extended by a DNA polymerase so that a copy is made of the target sequence. After making this copy, the same primers can be used again, not only to make another copy of the input sequence but also of the short copy made in the first round of synthesis. This leads to logarithmic amplification. For RNA viruses the RNA must be copied to DNA first using a reverse transcriptase (RT) step.

TMA is an isothermal method that allows exponential amplification of RNA or DNA targets in, as for PCR, the genome of the infectious agents being screened for. TMA technology uses two primers, one of which contains a promoter sequence for RNA polymerase, and two enzymes: RNA polymerase and reverse transcriptase. In the first step of amplification, the primer with the promoter sequence hybridizes to the target rRNA at a defined site. Reverse transcriptase creates a DNA copy of the target rRNA by extension from the end of the promoter primer. The RNA in the resulting RNA:DNA duplex is degraded by the RNase activity of the reverse transcriptase. Then a second primer binds to the DNA copy and a new strand of DNA is synthesised from the end of this primer by reverse transcriptase, creating a double-stranded DNA molecule. The RNA polymerase recognises the promoter sequence in the DNA template and initiates transcription. Each of the newly synthesised RNA, amplicons reenters the TMA process and serves as a template for a new round of replication. Again, as in PCR, this leads to logarithmic amplification.

However, molecular screening is a still emerging approach, not so much the underlying technology itself, which has been in use in research and development activities for over 20 years, but it is only recently that this technology has been successfully developed into the fully automated mass screening systems that include both the sample preparation and molecular aspects and which can provide results in just 3–4 h. However, although not to the same degree, sample quality is still a factor to be considered. Although designed to work with a range of different sample types, the initial sample handling side of the automated systems may, in the same way as with serological systems, reject unsuitable samples, those that are not detected or are rejected by the sample detection systems. However, molecular techniques do have a distinct advantage over serology in that each sample has an internal control added prior to the nucleic acid extraction process that is extracted and amplified concurrently with any specific target that may be present, and which is then detected using a separate detection system. Thus every sample should score as positive with the internal control, irrespective of the presence of the specific target of interest, uniquely, validating each negative result. Failure of the internal control indicates a failure of the process at some point, whether it be due to a failure of the molecular reagents themselves or due to inhibition of the reagents by something within the sample itself, and invalidates the results for the sample.

As discussed previously the value of molecular screening is primarily to detect early infection in those donors who have been recently infected. Later in infection the serological markers present are far easier to demonstrate and the assays significantly less expensive than molecular assays. The important question is then to determine the increased value provided by including molecular screening in addition to serological screening. Although molecular screening does theoretically offer earlier detection of infection, its actual value is very much related to the biology of the individual infectious agents, the overall performance of the serological screening and the incidence of infection in the population. Where antigen assays are available, there is generally only limited additional value to be had from molecular screening, unless the incidence of infection is high enough to indicate that significant number of donors may have been exposed to the agent. However, in low-risk populations this is rarely the case. However, where only antibody assays are available and where the window period of infection is relatively long, molecular screening may have more value. However, again it is the incidence of the infection in the donor population that determines how many donors would be likely to be detected by molecular and not by serological screening. In low risk populations the actual numbers of donors who would be detectable by molecular screening and not by serological screening is extremely low. Currently, for HCV in blood donations collected within the NBS, the number of HCV RNA positive but serology negative donations is approximately 1 in 55.4 million donations, i.e. one donation per 28 years at the current donation level.

Residual Risk

The overall effectiveness of the serological screening of tissue donations, both surgical and deceased, is without question. For many years serological screening was the only approach available for the screening of tissue donations and there is no published evidence, excepting in cases of haemodilution [10], that this approach has led to the transmission of infection as a result of serological testing failing to detect the presence of serological markers in an infected individual. The inclusion of molecular testing, at least theoretically, increases the level of safety even more, although the incremental benefit over that gained from serological screening alone is difficult, if not impossible, to quantify accurately. This is due to the relatively small numbers of tissue donations screened, the even smaller numbers of confirmed positive donors and the virtual impossibility of accurately determining the incidence of infections in tissue donors. Overall the microbiological safety of tissue donations that have been properly and effectively screened for infectious diseases is at least equal to that of blood donations.

Unfortunately however, laboratory testing is not perfect and there are occasions (for a number of reasons) when a test may not detect its target. Consequently no screening programme has absolute sensitivity and it is possible that infected donors could be undetected with the resultant entry of an infected donation into the supply chain. This is referred to as **residual risk**, i.e. the risk of infection from a screen negative donation. However, this is generally due to the lack of detectable target, rather than the assay failing to detect target.

In brief, assays fail to detect target either because the assay has missed something that is present or because there is nothing present for the assay to detect, although the donor is infected and the donation is infectious. The first scenario arises either due to poor assay performance (analytical sensitivity) or due to poor performance of the assay (operator/system error). However, if well evaluated and validated highly sensitive assays are being used, and the screening is being performed within a formal and well designed quality management system with well trained and competent staff, then this scenario should rarely, if ever, occur. The second scenario, lack of detectable target yet with infectivity present is, however, more likely to occur, no matter how good the screening programme and quality management system, as it is related to the natural history of infection in an individual and is a stage of the normal infection process.

Window Period

The major threat, in terms of residual risk is therefore from donors who are infected yet negative on screening as the normal screening target(s) are not present, or are present but below detectable levels. This situation occurs naturally very early during the course of infection, at a time when a productive infection has arisen, but before the specific screening target(s) have either been produced or their levels have risen to those detectable by the assays in use. It also occurs later in some infections when the infection is starting to resolve, the level of the screening target has declined to below that detectable, yet infectivity remains. This is referred to as the "Window period" of an infection, infectious yet undetectable with the screening assays used.

In relative terms, specific antibody appears later during the course of infection than antigen, and antigen later than nucleic acid. Therefore if the screening target is just antibody, there will be a period of time when donations from infected donors would give negative screening results despite the presence of antigen and its associated infectivity. The same applies when comparing antigen (with or without antibody) screening with molecular screening. In this situation the relative time between the first appearance of nucleic acid and the first appearance of antigen is much shorter, and in many instances they appear so close together that there is no actual measurable incremental benefit to be obtained through the use of molecular screening in addition to antigen or antigen/antibody screening.

The overall risk of collecting a donation from a donor who may be in the window period of infection is very much related to the overall level of infection in the population, and specifically the incidence of infection in the donor population. The higher the incidence of infection, the greater the risk that a donor could have been recently infected. At the same time the ability to detect an infected donation is dependent upon having an effective and appropriate screening programme in place. A screening programme with a relatively poor sensitivity is more likely to fail to detect recently infected donors. However the overall risk is also dependent upon the actual number of donations collected, and although the incidence of infection may be the same in the blood and tissue donor populations, as the populations are so very similar, the overall numbers of tissue donations collected is many orders of magnitude lower than the number of blood donations. Thus it can be argued that the overall probability of actually encountering a donation from a tissue donor who is in the window period, assuming a screening programme with equal if not greater sensitivity, is much lower than that for a blood donation. However, when the actual number of infectious identified in the UK are looked at the incidence of confirmed infected deceased tissue donors is significantly higher than that of blood donors [14], but the denominator is around 3 orders of magnitude less and therefore the figures cannot be compared directly. Furthermore significantly more patients have transfusions, greater number of potential exposures, and it is therefore not necessarily true to argue that tissues are a higher "risk" than blood or other donations.

The window period is thus the main reason today that donations, be they tissues or blood, have an associated residual (albeit very low) risk, and screening programmes and strategies are developed to ensure that the window period is reduced to as short a period as possible.

Conclusions

Overall the current screening of tissue donations prior to release is highly effective at preventing transmitted infections. The basic principles applied to the screening of blood donations do apply as there is considerable overlap in the need to identify low risk donors, the range of transmissible infectious diseases, the screening assays used and the screening strategies developed. Certainly there is little, if any, difference between the screening of surgical tissue donations and blood donations, as the samples are all collected by standard venepuncture from living individuals. The screening of deceased tissue donations is different inasmuch as the nature of the samples is fundamentally different, although the screening assay themselves are the same. It is the reliability of the screening process in relation to deceased samples that needs to be assured, particularly the representativeness of the samples collected in respect of the screening results obtained reflecting the true status of the donor and therefore the donation. In addition consideration should be given to the specific screening algorithms used, as those commonly used for blood donation screening may not always offer the most appropriate approach for the screening of deceased tissue donations. The issue of specificity is something that has particular relevance as the unnecessary loss of tissue donations due to poor specificity is unacceptable because tissue products are generally in shorter supply, often with many potential recipients on waiting lists. Furthermore a cadaver may provide a large number of different tissues for clinical use, and a non-specific screening result would result in the loss of a significant number of clinically valuable products.

References

1. Scardino MK, Hwang SJ, Hanna CL, Danneffel-Mandelkorn MB, Wilhelmus KR (2002) The postmortem sociomedical interview: uncertainty in confirming infectious disease risks of young tattooed donors. Cornea 21:798–802
2. Fischer SA, Graham MB, Kuehnert MJ, Kotton CN, Srinivasan A, Marty FM, Comer JA, Guarner J, Paddock CD, DeMeo DL, Shieh W-J, Erickson BR, Bandy U, DeMaria A, Davis JP, Delmonico FL, Pavlin B, Likos A, Vincent MJ, Sealy TK, Goldsmith CS, Jernigan DB, Rollin PE, Packard MM, Patel M, Rowland C, Helfand RF, Nichol ST, Fishman JA, Ksiazek T, Zaki SR, and the LCMV in Transplant Recipients Investigation Team (2006) Transmission of lymphocytic choriomeningitis virus by organ transplantation. NEJM 354:2235–2249
3. Cahane M, Barak A, Goller O, Avni I (2000) The incidence of hepatitis C virus positive serological test results among cornea donors. Cell Tissue Bank 1:81–85
4. Challine D, Roudot-Thoraval F, Sabatier P, Dubernet F, Larderie P, Rigot P, Pawlotsky JM (2006) Serological viral testing of cadaveric cornea donors. Transplantation 82:788–793
5. Heim A, Wagner D, Rothamel T, Hartmann U, Flik J, Verhagen W (1999) Evaluation of serological screening of cadaveric sera for donor selection for cornea transplantation. J Med Virol 58:291–295
6. Padley D, Ferguson M, Warwick RM, Womack C, Lucas SB, Saldanha J (2005) Challenges in the testing of non-heart-beating cadavers for viral markers: implications for the safety of tissue donors. Cell Tissue Bank 6:171–179
7. Thomas S, Klapper PE, Mutton KJ, Turner AJ, Tullo AB, Zambrano I, Carley F, Taylor A (2007) Lack of vision, loss of sight: consequences of mandatory HTLV screening in corneal transplantation. Transpl Infect Dis 9:171–172
8. Miedouge M, Chatelut M, Mansuy JM, Rostaing L, Malecaze F, Sandres-Saune K, Boudet F, Puel J, Abbal M, Izopet J (2002) Screening of blood from potential organ and cornea donors for viruses. J Med Virol 66:571–575
9. Eastland T (2000) Hemodilution due to blood loss and transfusion and the reliability of cadaver tissue donor infectious disease testing. Cell Tissue Bank 1:121–127
10. Centres for Disease Control (1987) Human immunodeficiency virus infection transmitted from an organ donor screened for HIV antibody. MMWR 36:306–308
11. http://www.fda.gov/BiologicsBloodVaccines/GuidanceComplianceRegulatoryInformation/Guidances/Tissue/ucm073964.htm#THEDONOR-ELIGIBILITYDETERMINATION1271.50

12. Kitchen AD, Gillan HL (2010) The serological screening of deceased tissue donors within the English Blood Service for infectious agents – a review of current outcomes and a more effective strategy for the future. Vox Sang 98:e193–e200
13. Kitchen AD, Newham JA (2010) Qualification of serological infectious disease assays for the screening of samples from deceased tissue donors. Cell and Tissue Banking (in press)
14. NHSBT Infection Surveillance Programme Annual Report 2007. http://www.hpa.org.uk/web/HPAwebFile/HPAweb_C/1227255714122

Chapter 11
Testing the Tissue and the Environment

Ted Eastlund and Martell K. Winters

Introduction

Testing for infectious microbes such as bacteria and fungi (yeast and molds) plays an important role in ensuring the safety of tissue allograft transplantation. Testing is routinely performed on the donated tissue prior to processing, on the allograft after processing as final testing, and to prevent contamination and cross contamination during processing. In the past, bacterial and fungal infections transmitted from contaminated tissue allografts were considered rare anomalies, whereas donor-derived HIV, hepatitis and other viral and prion infections were considered more important threats to patients. Indeed, tissue transplantation has resulted in donor-to-recipient transmission of HIV, hepatitis, human T-cell lymphotropic virus, rabies, *Herpes simplex*, Epstein-Barr virus, cytomegalovirus, and transmissible spongiform encephalopathies [1, 2]. Tissue banks and standard-setting bodies have emphasized the importance of preventing these viral and prion diseases, even though the risk is low. In most countries, blood testing of tissue donors for infectious diseases parallels that for blood donors and includes antibody testing for viruses and syphilis and when needed for selected parasites. In some countries, direct blood testing for HCV and HIV nucleic acid has been implemented for tissue donors as well as with blood donors. In contrast with blood component transfusion, testing for tissue allografts also relies on the routine use of microbiologic testing for bacteria and fungi to ensure patient safety, as explained in this chapter.

Bacterial and fungal contamination of tissue allografts obtained from cadaveric donors has been a serious cause of morbidity and mortality in many recipients of bone, cornea, skin, heart valve, tendon, pericardial and cartilage allografts [1, 3]. Most of these infections could have been easily prevented. Until recently, the threat to patients of receiving a tissue allograft contaminated by bacteria or fungi has been thought to be minimal but several recently reported fatal and nonfatal cases have drawn attention to this problem [3, 4].

T. Eastlund (✉)
Department of Pathology, University of New Mexico School of Medicine,
Albuquerque, NM 87131, USA
e-mail: deastlund@salud.unm.edu

G. Galea (ed.), *Essentials of Tissue Banking*, DOI 10.1007/978-90-481-9142-0_11,

Table 11.1 depicts the published sources of contamination by bacteria and fungi that caused infections in allograft recipients [4–23]. These sources include undiagnosed bacteremia and sepsis as a cause of death of the donor, contamination of the tissue by endogenous bacteria and fungi shortly after death, contamination from the processing room environment, contaminated processing reagents, erroneously omitting sterilization steps, and the use of insensitive unvalidated final sterility tests for bacteria and fungi prior to packaging. Consequently, these cases have called attention to the importance of accurately diagnosing and excluding tissue donors with bacterial sepsis, avoiding processing of donated tissue carrying high loads of virulent bacteria, using carefully validated disinfection and sterilization steps during tissue processing, following trends of clean room environmental monitoring results and setting action levels, and the use of validated final sterility tests and bacteriostasis and fungistasis testing when antibiotics are used during processing.

In addition, these cases show that a terminal sterilization step is important to use whenever possible, particularly for allografts that are able to withstand sterilization without loss of effectiveness. With use of carefully validated sterilization methods

Table 11.1 Published bacterial and fungal transmission from tissue allografts

Contamination source	Allograft	Microbe (references)
Septic donor, microbe on recovered tissue	Tendon	*Streptococcus pyogenes,* Grp A [5]
Known infected living donor	Bone	*M. tuberculosis* [6]
Uninfected donor, microbe on recovered tissue	Heart valve	*Candida albicans* [7]
Uninfected donor, microbe on recovered tissue	Cornea	Bacteria and fungi [8–11]
Uninfected donor, microbe on recovered tissue	Bone	Several bacteria [12, 13]
Uninfected donor, microbe on recovered tissue	Cartilage	*Clostridium sordellii* [14–16]
Uninfected donor, microbe on recovered tissue	Tendon	*Clostridium septicum* [15, 16]
Uninfected donor, microbe on recovered tissue	Skin	*Pseudomonas aeruginosa* [17]
Processing room environment	Tendon	*Elizabethkingia meningiosepticum* [4]
Recovery/processing room	Heart valve	Oerskovia turbata [22, 23]
Processing reagent fluid	Pericardium	*Ochrobactrum anthropi* [19–21]
Inadequate disinfection during processing	Several	Several [4, 5, 7–11, 14–17]
Human error: terminal irradiation omitted	Tendon	Several bacteria [18]
Falsely-negative final pre-packaging culture	Cartilage	*Clostridium sordellii* [14–16]
Falsely-negative final pre-packaging culture	Tendon	Several bacteria [4, 5, 15, 16]
Falsely-negative final pre-packaging culture	Heart valve	*C albicans* [7]

for tissues, tissue banks gain the added benefit of possibly moving away from final sterility testing of each tissue; as long as emphasis shifts from reliance on final tissue testing to in-process controls at key points in the processing stream. This chapter addresses issues surrounding microbial testing at various critical points, including recovered tissue prior to processing, the tissue processing environment and the final tissue allograft itself.

Microbal Testing of Tissues Recovered Prior to Processing

Testing Tissue from Deceased Donors

Microbial testing of recovered tissues before processing is important because the same microbes causing infections in recipients of bone, tendon, cornea, skin and heart valve allografts were present from samples obtained immediately after recovering the tissue from the donor [2, 5, 7, 12, 13, 17]. Pre-processing microbial testing is routinely performed by swabbing each piece of recovered tissue, placing the swab into a transport medium, transporting the swab to the test laboratory, placing the swab into a growth medium and incubation of the growth medium. Usually this testing is performed to detect both aerobic and anaerobic bacteria and fungi. This testing is required by some standard setting organizations [24]. Tissue banks establish acceptability criteria for the results of recovery cultures to determine whether the tissue will be processed and sometimes to determine which processing steps will be used.

Microbial contamination of recovered tissues is often detected even though the donor has no evidence of infection while alive. Swabbing is the most common method for testing recovered bone and tendons and bacteria of low and high virulence are commonly found. Tissue banks report that 4–53% of recovered tissue has bacterial or fungal growth [13, 25–44]. Most microbes were low virulent ones such as coagulase negative staphylococcus. Tissue banks that recover hearts for producing heart valves sample the transport media by filtering it prior to processing and the filter is then incubated thus increasing the sensitivity of detection. Pieces of recovered skin are commonly tested by an immersion method. Depending on the type of allograft, the finding of certain microbes can be cause for discarding the tissue or determining what type of processing to perform.

Microbial contamination of donated tissues can arise not only from the recovery site environment, from donor skin [40] and recovery staff during tissue recovery but also because of the expected postmortem spread of intestinal microbes from intra-luminal to extra-luminal sites through lymphatic and vascular structures as part of the normal postmortem decomposition of the body. The epithelial mucous membrane of the intestine is a very fragile barrier protecting a person from the intestinal bacteria. This barrier breaks down easily after a person's death and even after short periods of hypotension or ischemia in living persons. This dissemination of endogenous microbes is called bacterial translocation [45–47]. It is likely that tissues recovered in deceased donors become contaminated in part through this mechanism.

In an attempt to limit this source of endogenous microbial contamination, the collection of tissue allografts from the cadaveric donor should begin as soon after death as possible and the body should be refrigerated. Removal of tissue allografts from cadaveric donors is completed by most tissue banks within 24 h of the death of a cadaveric donor. Some professional standards require bone, tendon and heart valve recovery to be performed within 12 h of death if the body is not refrigerated and within 24 h of death if the body is refrigerated [24]. In some countries tissue banks undertake tissue recovery up to 48 h after death of the donor. Delayed removal of tissue from the cadaveric donor can possibly result in an increased contamination by anaerobic and spore forming bacterial pathogens which are more resistant to disinfecting and sterilization procedures. Although there is some evidence for its support, it is unknown whether the risk of post-mortem bacterial translocation increases when donors have died from trauma, gunshot injuries or have undergone prolonged cardiopulmonary resuscitative attempts, autopsies, and other post-mortem physical manipulation.

The environment in which the tissue allograft is excised from the donor can also have an impact on microbial contamination. Bacterial contamination can be minimized by ensuring that the recovery site is as clean as possible. Cleanliness of the recovery site can be controlled by adopting recovery site suitability requirements [48]. Tissues are removed at various recovery locations: hospital operating rooms, autopsy suites in hospital morgues and regional forensic medical examiner facilities, funeral homes (mortuaries), and in dedicated tissue procurement facilities at medical examiner facilities or tissue banks. Although not a consistent finding by others, one study reported that prior organ donation and the location of the procurement site are correlated with finding virulent bacteria on tissue allografts removed from cadaveric donors [49]. They found virulent bacteria on at least one tissue in 58 (21%) of 275 donors and in 125 (3.5%) of 3,586 individual tissues. In contrast, non-pathogenic bacteria were found on 1,622 (17.3%) of tissues.

Testing of Tissue from Living Donors

The most common tissues donated by living donors are femoral heads that would otherwise be discarded during total hip arthroplasty surgery. Bacterial growth on the surface of bone removed from living tissue donors is found less often than on tissues removed from deceased tissue donors. Bacteria have been found on 3.4–22% of femoral heads donated by patients undergoing total hip arthroplasty surgery [50–56]. Most reported contamination rates are 3.4–5.7% with common low virulence bacteria consistent with contamination from air and skin [50–54].

Postmortem Blood Cultures

Blood samples from deceased tissue donors have often been obtained at the time tissue is recovered and tested for bacterial and fungal growth. Positive postmortem blood culture results have not been very predictive of infection as a cause of death or predictive that recovered tissues will be contaminated by the same microbe. Consequently postmortem blood cultures are not uniformly performed or required by professional standards or governmental regulations.

Although the donors are not clinically infected at the time of death, postmortem blood cultures are frequently positive with rates as high as 23% [32], 32% [25] and 39% [57]. The majority of the isolates are low virulence bacteria, most often skin contaminants. Coors et al. [57] showed that 13.8% were positive for virulent bacteria, 25.6% positive for low virulence bacteria and that half of the positive blood cultures were from skin contaminants acquired at the time of sampling. They also found that the rate of positive blood cultures of deceased tissue donors taken many hours after death was higher than that of deceased organ donors whose circulation was intact and that the positivity rate was higher when the time after death was greater.

Environmental Monitoring of Cleanrooms Used for Tissue Processing

Severe meningitis in children, septic bacterial arthritis, and endocarditis with bacteremia have been caused by allografts contaminated by bacteria acquired during tissue processing from contaminated washing fluids and contaminated locations within the cleanroom environment [5, 19–23]. The causative environmental bacteria were *Ochrobactrum anthropi, Oerskovia turbata and Elizabethkingia meningeosepticum.* These cases illustrate the importance of microbial control in cleanrooms.

Tight control over the operation of the cleanroom used for aseptic processing of tissue and controlling incoming contamination with incoming materials and staff is the primary safeguard against bacterial contamination of tissues during processing. Careful aseptic technique, use of sterile equipment, supplies and reagents, routine cleaning after use and microbial monitoring all contribute to preventing allograft contamination during processing in a cleanroom. Monitoring for the presence of microbes in clean rooms is also extremely important by documenting the presence and concentration of bacteria and fungi in the air and on smooth surfaces. Other sources of contamination to monitor are walls and other surfaces with irregularities, seams and junctions or thick residues and moist water traps that can retain microbes after routine cleaning and application of sanitizers, disinfectants and sporicides.

Cleanroom Classification

There are many factors that increase the chance of tissue becoming contaminated during processing (Table 11.2). The use of cleanrooms for tissue processing is one of the most important precautions for control of microbial contamination. Cleanrooms are validated and classified based primarily on the cleanliness of the air which is being used in the room. This cleanliness is estimated by the concentration of airborne particles. Most particles are nonviable but viable particulates represent living microbes, bacteria and fungi (yeast and molds). Table 11.3 depicts the classification of cleanrooms based on total particulate levels, including both viable and nonviable particulates [58].

Table 11.2 Factors influencing microbial contamination during processing

Microbes in the air
Microbes on surfaces, drains
Microbes shed from staff and equipment brought into room
Contaminated solutions, reagents, rinses
Contaminated surgical supplies, equipment
Inadequate cleaning, sanitizing and disinfecting of the cleanroom
Inadequate tissue disinfection steps
Inadequate tissue sterilization steps
Errors during processing or sterilization
Insensitive sampling for final sterility testing
Residues of processing antibiotics interfering with final sterility testing

Table 11.3 Cleanroom classification based on allowable airborne particulate levels

Class	≥0.5 μm particles/m^3	Microbiological active air, action levels (cfu/m^3)	Microbiological settling plates, action levels (90 mm diameter, CFU/4 hours)	Approximate EU grade	209E equivalent
ISO 2	4	0	0		N/A
ISO 3	35	0	0		Class 1
ISO 4	352	0	0		Class 10
ISO 5	3,520	1	1	A, B	Class 100
ISO 6	35,200	7	3		Class 1,000
ISO 7	352,000	10	5	C	Class 10,000
ISO 8	3,520,000	100	50	D	Class 100,000
ISO 9	35,200,000	N/A	N/A		Room air

This table was compiled from the following standards: (1) ISO 14644-1; (2) Revision of the Annex to the European Union (EU) Guide to Good Manufacturing Practice-Manufacture of Sterile Medicinal Products, Jan 1997; (3) United States Pharmacopeia (USP) <1116>; (4) United States FDA, Guidance for Industry, Sterile Drug Products Produced by Aseptic Processing – CGMP, Sep 2004; (5) Previous United States Federal Standard 209E

Standards, best practices and industry trends create expectations as to the level of cleanliness that is required for its use. It is common to have tissue processing cleanrooms follow the same standards as medical device and pharmaceutical companies. Cleanrooms classified as Grade A or ISO Class 5 (previously known as Class 100) or EU Grade A or B are common for tissue processing in many areas of the world. Cleanroom operating standards are found in United States Pharmacopeia (USP), European Pharmacopeia, and International Organization for Standardization (ISO). Regular testing and monitoring of airborne particulate concentrations is necessary to ensure that requirements of the cleanroom standard are maintained.

Sampling of Cleanroom Air for Viable Particulates

In addition to air sampling for particulates to ensure that overall cleanliness and air quality are maintained, sampling for viable particulates (microbes) which can contaminate tissue allografts during processing is important. Both specific types of microbes and concentrations of each type should be monitored. Results should be followed to establish acceptable baseline levels and identify excursions above these baselines. Airborne microbial counts are usually sampled by two main types of devices; passive and active air sampling devices. Air samples provide an average number of colony forming units (CFU) per unit of time or CFU per liter of air sampled. This is called a CFU because what grew into a visible colony could have started as an individual microorganism or a clump of microorganisms.

Passive Air Sampling: Settling Plates

Passive air sampling involves an agar plate left opened on a surface somewhere in the cleanroom for a certain period of time, often 30–60 min or up to several hours. Microbes settle on the agar and grow when incubated and colonies are counted. If the plates are left open too long they may begin to dry and not support growth. Settling plates are most commonly used, require little training, inexpensive to perform and easy to use but do not supply quantitative data as CFU per measured air volume.

Active Air Sampling

In an active air sample, equipment of one type or another draws air from the cleanroom at a certain rate and passes the air by agar in a plate or strip. Air sampling devices have varying degrees of efficiency at capturing viable microorganisms. Centrifugal samplers spin an aerosol at a high rate of speed and rely on centrifugal force to propel particles against a settling plate. Filtration devices collect viable particles by impelling air against a filter, such as gelatin membranes or cellulose fibers, that can then be placed onto agar for enumeration. Impingement devices entrap microbes in a liquid medium as air is transmitted through the fluid. Impaction samplers use a vacuum to draw microbes onto an agar surface.

Results are usually provided in CFU per liter, per cubic foot or per cubic meter of air. The benefit to this approach is that the air can be drawn from a very specific location in the cleanroom. For example, air can be drawn from a specific HEPA filter or adjacent to a piece of equipment which causes turbulence in the air flow (e.g. a bandsaw).

Sampling of Cleanroom Surfaces for Microbes: Contact Plates

Surface sampling can be performed using contact or "RODAC" plates (a trademark name for Replicate Organism Detection And Counting) which are Petri dishes containing a sterile agar growth medium that protrudes above the sides of the plate so that it can be touched to flat, regular surfaces. Microbes on that surface will adhere to the agar and grow upon proper incubation providing a quantitative result of the number of CFU on the surface. When RODAC samples are taken in a class 100 (grade A, ISO 5) cleanroom, it is common to set an alert level at 1–2 CFUs. In class 10,000 (grade C, ISO 7) areas, the alert levels are usually 5–10 CFUs and in class 100,000 (Grade D, ISO 8) areas, most have set the alert levels at 25–50 CFUs.

Contact plates are easy to use and are widely available but may not be appropriate for irregular surfaces, where swabbing will be needed. If disinfection of the surface has recently occurred it is critical to assure that the disinfectant has completely dried prior to performing the sampling. The agar used in this testing often has a neutralizer mixed in to counter any residual disinfectant which may be on the surface being sampled.

Sampling of Cleanroom Surfaces for Microbes: Swabbing

Swabbing is performed to obtain surface microbial counts from irregular contoured surfaces as well as tubing, equipment and drains. After sampling, the swab is either vortexed in a tube of sterile fluid followed by testing of the fluid or the swab is simply streaked onto an agar plate after swabbing the surface. Of these two, the former is the better approach but does require additional equipment and training.

Sampling Locations

No standard will specify exactly which areas of a cleanroom should be monitored for viable microbial counts. Generally common sense will dictate the critical areas of the cleanroom and those are where the monitoring should occur. Normally monitoring should be performed in all areas where there is direct interaction with the allograft tissue, at least initially. For example, if the allograft rests on a table top for some time it would be appropriate to regularly monitor that area for surface contamination.

It is better to sample an excess number of areas during initial qualification of the room or when implementing a new processing step until enough data is obtained to determine which of the sites are the best indication of the general cleanliness of the room. At that point it may be appropriate to reduce the quantity of sites to something more reasonable for the long term.

Sampling Timing

There are two high-risk time periods for contamination to be acquired in the clean-room environment and air and surface sampling for microbes should be performed at times to consider these. Most important is the time period between when the cleanroon is cleaned and disinfected and the onset of the active processing of tissues by staff (immediately pre-processing). Staff entering the just-cleaned room can riddle it with microbes and particulates that can shed from their entry and from the supplies, equipment and tissues they bring into the room. These microbes are then present during tissue processing. The second time is after processing has finished and before the room is cleaned (immediately post-processing). Breaches in aseptic technique can occur when processing is complete and staff are less conscious of cleanliness.

Trend Analysis and Action Levels

In a cleanroom environmental monitoring program, isolates should be identified to the genus level and if possible to the species level. The types and concentrations of isolates are recorded and trends examined and identified. The routine review of cleanroom microbial test results can identify virulent pathogens or a trend towards increasing levels of contamination that may require corrective action. Action levels are set to trigger a planned set of corrective actions to return the processing room to acceptable limits as quickly as possible. Staff can promptly intervene when values outside acceptable levels are detected to diagnose root causes and take preventive action in order to ensure tissue allograft safety and to avoid harm to patients and economic loss to the tissue bank. Exceeding the alert levels should also increase sampling frequency in the affected area.

Final Microbial Testing of Allografts After Processing

Final Allograft Testing

Fatal and nonfatal bacterial and fungal infections were caused in recipients of cartilage, tendon and heart valve allografts that tested negative in final testing but nervertheless were contaminated [5, 7, 14–16]. These cases emphasize the importance of sensitive reliable final testing. Microbial testing of fully processed tissue immediately prior to final packaging, often called final sterility testing, can be performed on each allograft, each co-processed tissue or a predetermined quantity of tissue at the completion of processing each batch depending of the type of tissue processed and type of disinfection or sterilization steps used. This testing is performed to detect the growth of aerobic and anaerobic bacteria and fungi (yeasts and molds). Cardiovascular tissue is usually also tested for *M. tuberculosis.* As with other critical steps in tissue banking, final testing should be validated to ensure the method reliably accomplishes that which is intended, including attention paid to sampling method and bacteriostasis and fungistasis testing. If the allograft is not terminally sterilized, each should be tested at the time of final packaging.

The purpose of final testing of the allograft is usually to demonstrate that the individual tissue being tested is free of contamination by bacteria and fungi. Since it is not a test for viral contamination, it cannot be accurately claimed to be a sterility test which implies complete absence of viable microorganisms, including viruses. Final microbial testing of the tissue allograft by itself is sometimes incorrectly assumed to indicate that an entire batch or lot is free of microbial contamination or has been sterilized. Final microbial testing alone cannot achieve this. Instead, this is accomplished primarily by validation of the entire aseptic processing procedure or sterilization process providing evidence that the largest expected bioburden can be reliably sterilized with a SAL of 10^{-6}. Statistically it is usually not possible to extrapolate the sterility of an entire batch by testing a few samples from that batch.

One shortcoming of final microbial testing includes falsely negative results that have caused many cases of fatal and nonfatal transmission of bacteria and yeast through cartilage, tendon and bone allograft use (Table 11.1). These false negative results may be due to faulty sampling but more likely were due to inhibition of microbial growth, usually by residual processing antibiotics, which must be overcome in final testing. When final testing shows microbial growth, there should be an investigation to look for failure of one or more of the tissue processing steps. Clusters of positive tests with the same organism are especially important. In one investigation, the positive results from final testing of several allografts was due to use of contaminated water during processing [59].

Sampling Methods – Swab Sampling

Historically swabbing methods have been extensively used in the tissue banking profession, mostly as a qualitative method for determining the presence or absence of bacteria and fungi on tissue. Swab sampling is used for recovered tissue, for environmental monitoring and for final testing. The advantages of swabbing are that it is inexpensive to perform the test, requires minimal training and equipment and that it has a long history of use. Its disadvantages are that the results can be quite variable depending on the technician performing the swabbing, the type of tissue being tested, the typically low efficiency of recovering microorganisms and that for most uses of swab testing, the data provides only the presence or absence of recovered microorganisms (qualitative) rather than a quantitative result.

Bacterial and fungal testing by swabbing of the bone and tendon tissue surfaces is routinely performed by many tissue banks at the time of tissue recovery. The detection rate of bacterial contamination by swab culturing of the entire allograft surface is much lower than if the entire bone had been immersed in culture media [30, 38]. The sensitivity of a swab method depends largely on the skill of the technician, the percentage of the tissue which is swabbed, and the technique used to remove the microbes from the swab for testing in the growth medium. These variables are difficult to control consistently because they are largely human-based.

In the context of routine microbial surveillance of recovered tissue prior to processing, swab testing, a non quantitative test, provides no assurance that incoming contamination has not exceeded the capacity of a validated tissue disinfection and sterilization process or will result in a safe allograft. Qualitative and quantitative testing and process surveillance is needed for this assurance. Qualitative swab testing of recovered incoming tissue has been used by tissue banks for other purposes, such as a guide for rejecting tissue with highly virulent organisms. This can be especially useful for allografts which are not able to undergo sufficiently strong, validated disinfection or sterilization steps that can harm the viability, structure and effectiveness of some allografts.

A properly validated tissue bank process does not exclude the need for good microbiological control of incoming tissue. In contrast, it requires it and validation provides an understanding of the ability of the process to remove or kill known quantities and types of microbes. Thus, during initial process validation and regularly thereafter, it is critical to document both the number and type of microorganisms present on the tissue rather than just the type. Many factors can lead to unpredictable bioburden of incoming recovered tissues (Table 11.4). Some studies have shown that incoming tissue bioburden can be large, exceeding 1.6 million CFU per bone allograft [60]. Swabbing lacks the capacity to determine if a positive result was due to a single microbe or one million microbes on the tissue.

On the other hand, it possible to perform a modified swab test of tissue and obtain an estimate of the number of microorganisms (quantitative) and validate its capacity. To do this one must determine the efficiency with which the swab picks up all the

Table 11.4 Factors limiting the reliability of microbial testing of recovered tissues prior to processing

Unpredictable bioburden due to postmortem translocation from intestine
Less that 100% efficiency of sampling method
Unpredictable staff performance in sampling
Sample storage and transport deficiencies
Sensitivity of microbial detection and identification testing
Swabs can miss high concentrations that can overwhelm the sterilization process

microbes encountered on the allograft, and the efficiency with which one is able to extract all the microbes from the swab.

To evaluate this, the tip of the swab is cut off and placed into a sterile extraction fluid, an extraction of the swab tip is performed (often some type of shaking or vortexing), the fluid is filtered through a membrane filter (usually 0.45 μm), the filter is placed onto an agar plate and the plate is incubated for growth. After incubation the microorganisms which have grown are counted and documented as colony forming units (CFUs). Even though this method cannot provide an exact number, it can provide a good estimation of the initial number of microbes present on the tissue. This method does require more equipment than a qualitative swab method in that a membrane filtration system must be obtained. This method, and its validation, are explained in ISO 11737-1 [61].

Swab Efficiency Validation

Swab testing generally has a low recovery efficiency. As low as 5% of the microbes present on the tissue are captured on the swab, but efficiency values in the 80% range and above have also been seen. By calculating the efficiency and using it as a conversion factor, one can estimate bioburden from numbers obtained by swabbing. Validation of a quantitative swab method can be performed in one of two ways: exhaustive rinse or inoculated product. In evaluating the swab recovery efficiency using the exhaustive rinse method, a swab is first used on tissue then extracted several times. It is common to perform this extraction three to four times. After each extraction the extraction fluid is tested to determine the number and type of microorganisms, as described above. The quantity of microorganisms which were removed in the first extraction of the swab is compared to the total number of microorganisms which was removed in all of the extractions. An example follows:

Extraction 1: 23 CFU
Extraction 2: 15 CFU
Extraction 3: 8 CFU
Extraction 4: 1 CFU

The recovery efficiency is calculated as the number of CFU Recovered on Extraction 1 divided by the total CFU recovered for all extractions.

$$23/(23 + 15 + 8 + 1) = 23/47 = 0.489 \text{ or } 48.9\%$$

If performed using multiple technicians and over multiple days, operator bias can be reduced and a valid recovery efficiency percentage determined. This provides a value for the difficulty in removing microorganisms from the swab once they are on the swab. The other factor to determine is the difficulty in removing microorganisms from the tissue. To determine this value using the exhaustive rinse method, the same piece of tissue is swabbed multiple times, each time with a different swab, and then the swabs are individually tested for bioburden, again as described above.

One difficulty in use of swabs for bioburden testing is that it involves a compounded recovery efficiency situation. To obtain the correct recovery efficiency percentage for swabbing of tissue, the value for removing the microorganisms from the tissue and the value for removing the microorganisms from the swab are multiplied together. Thus, if the efficiency of removing the microorganisms from the tissue is 55.8% and the efficiency of removing the microorganisms from the swab is 48.9% (as above), the cumulative value is:

$$0.558 \times 0.489 = 0.273 \text{ or } 27.3\%$$

This recovery efficiency can be used to adjust the microorganism count obtained from a swab test of the tissue (e.g., when the bioburden count of 18 is found by swab testing) using the following method:

Swab Bioburden Count / Recovery Efficiency = Bioburden Estimate
18 CFU/0.273 = 65.9 CFU (66 CFU) on the swabbed allograft

When the naturally occurring bioburden of the recovered tissue is low, which is known to usually be the case, the inoculated product recovery efficiency method is recommended. In this approach a known quantity of microorganisms is added to the tissue, which is then swabbed, extracted, filtered, incubated and counted. In this validation method, however, it is not required to perform multiple extractions on the same piece of tissue or on the same swab because the quantity of microorganisms added to the tissue is known. If the natural tissue bioburden is known to be low, and if the correct microorganism is used for this test (one which is easily distinguishable from typically occurring microorganisms on the tissue such as *Bacillus atrophaeus*) it is usually not necessary that the tissue be sterile prior to the test. The calculation of recovery efficiency in this case is as follows:

CFU Recovered / CFU Inoculated onto Tissue
46 CFU/61 CFU = 0.754 or 75.4% efficiency

The inoculated allograft recovery efficiency estimated by the method can be used to adjust bioburden counts in the same way as the use of an efficiency value derived by the exhaustive rinse method.

Sampling Methods – Whole Allograft Immersion and Extraction (Bioburden Testing)

The whole immersion method usually provides a higher recovery efficiency than swabbing and potentially can be designed to allow for use of the tissue after the test. In this approach the tissue (a single piece or multiple pieces) is placed into a sterile container and extracted in a sterile solution (similar to the treatment of the tip of the swab described above). The solution is then filtered or tested in some other way, and then incubated and counted.

For medical devices the extraction solution commonly used is 0.1% peptone and 0.1% polysorbate 80 solution, also called Fluid D. These additives are used to assist the microorganisms in maintaining proper osmotic pressure and allow for good microbial growth (peptone) and to assist in removing the microorganisms from the tissue (polysorbate 80, which is a detergent). If the tissue is to be used clinically after the extraction, steps must be added to remove as much residue as possible and determine what impact potential residues of this type may have on allograft effectiveness and recipient safety. Alternatively, the extraction fluid can simply be sterile water or buffer.

The immersion and extraction method provides more consistent data than swabbing because the variables are more easily controlled. The extraction method is often a mechanical shake on a piece of equipment, which is not used in swabbing. This method is more equipment and training intensive than the swab methods.

One issue which sometimes must be considered with this method is the buildup of lipids extracted from the tissue and remaining in the extraction solution. This can make the solution difficult to filter through a 0.45 μm filter. Filtration of multiple aliquots of solution through multiple filters, or use of pour plate methods in multiple plates can be used if this is the case.

Sampling Methods – Destructive Testing

This testing usually involves immersing the tissue into a microbial growth medium and incubating for growth. At the end of incubation the test is scored as positive or negative for growth (a qualitative test). It is common for this test to incubate for 14 days and to be tested using a medium specific for aerobic growth (e.g. trypticase soy broth) and one for anaerobic growth (fluid thioglycollate medium). The test can be performed using either tissue that ordinarily would be used for implantation, or other tissue from the same donor, fully processed the same as the other tissue but which could not be used for implantation (companion tissue or co-processed scraps of tissue which are trimmed off of tissue to be implanted).

This test is very sensitive since the tissue itself is tested; there are no issues regarding the ability to remove the microorganisms from the tissue. The disadvantages are that the tissue must be discarded after this test and that it is qualitative, not quantitative. Testing using this method is commonly employed for cardiovascular tissue which cannot undergo sterilization steps because its harmful effects on viability and function.

Bacteriostasis/Fungistasis (B/F) Testing

Before relying on the use of a final test to determine whether a tissue is suitable for release, bacteriostasis and fungistasis testing should be performed to ensure that any bacteriostatic or fungistatic activity inherent to the allograft does not adversely affect the reliability of the test. B/F testing verifies that any negative results are due to the lack of viable microorganisms rather than being falsely-negative due to something on or in the allograft which is inhibiting microbes from growing, e.g, residual antibiotics or disinfectants used during processing. Typically this is demonstrated by adding dilute cultures of bacteria and fungi (e.g., *S. aureus, P. aeruginosa, C. sporogenes, C. albicans, A. niger, B. atrophaeus)* at less than 100 colony forming units per mL to a container with the media and the allograft, and demonstrating that those microorganisms can grow to a level where the media becomes turbid while in the presence of the allograft. [62, 63].

Tissue processing commonly includes exposure to alcohols, peroxides, detergents and antibiotics which can remain on the tissue as a residue. Following B/F testing, tissue banks commonly prevent the effect of possible inhibiting residues by diluting them with large predetermined volumes of culture media (amounts determined by B/F testing) in the final testing. Another method to reduce the possible effect of inhibitors is the use of additives in the sterility test media. Among the more common additives are polysorbate 80, lecithin and sodium thioglycollate.

Final Testing of Sterilized Allografts

Even though final sterility testing is commonly performed on representative samples of sterilized allografts, it may not be needed if sterilization validation was done properly. A properly validated sterilization process will usually provide at least a 10^{-6} sterility assurance level (SAL), which provides a one in one million probability of finding a non-sterile tissue after the sterilization process. Using final sterility testing to document that the sterilization process has worked properly would require testing of one million pieces of tissue after sterilization. If only ten samples are tested after sterilization and found negative, this data supports a mere 10^{-1} SAL estimate, which demonstrates a one in ten probability of a non-sterile tissue. Bridging the gap between the demonstrated 10^{-1} SAL and the desired 10^{-6} SAL by using more final testing would require a million allografts to be tested.

Testing of allografts after they have undergone sterilization steps usually suggests that appropriate sterilization validation has not been performed, and there is a concern for the capacity of the sterilization process to properly sterilize the tissue. Following sterilization validation, final sterility testing can be replaced by controlling the validated process. For example it requires use of indicators to ensure that sterilization was applied and at a sufficient dose to achieve the desired SAL. This is often either a spore strip (a biological indicator) for most sterilization types or a dosimeter demonstrating that the proper dose of radiation has been received. Biological indicators, dosimeters, controlled bioburden on incoming tissue, and verification that the sterilization process met specifications can assure that the process was carried out as validated and that the selected SAL was achieved, even without final sterility testing.

Process Validation and Sterility Assurance

Historically, it has been common for tissue banks to develop a tissue processing method based on another tissue bank's process with a good safety record and demonstrated good function rather than by analyzing microbiological requirements throughout the process. It was common to equate an effective process with the results of the final tissue sterility test. A passing final sterility test was meant to indicate that the entire process was under control and that the entire lot or batch of the tissue was also sterile.

An improved approach taken by many tissue banks at developing a tissue process that results in safe tissue allografts has been to inoculate tissue at the beginning of the process with one or several types of microorganisms at high titers, put the tissue through the entire process, and test for microbial growth at the end of the process. Although this approach can provide some microbiological information regarding the overall process, it cannot be termed true "process validation".

More recently, validation approaches which document the predictability and reproducibility of the microbial reduction capability of the process are more common. In these approaches each step of the process is evaluated separately, such as debridement and washing and the stepwise application of various disinfectants and sterilants. Each of these sequential processing steps are evaluated from a microbiological reduction (log kill) standpoint, or are evaluated regarding effects on maintaining tissue appearance, form and function.

Data from validation testing can demonstrate how to optimize a processing step from a microbiological standpoint and a tissue function standpoint. Alternatively, depending on the data, a step in the process may even be eliminated if it does not provide sufficient benefit to the tissue. This stepwise validation analysis can assist in determining appropriate processing ranges for variables such as time, temperature or solution concentration. Process validation of this type can not only be valuable in establishing the best processing method for tissue but also later for evaluating excursions from normality which inevitably arise in the course of a tissue bank's life. Although proper process validation can be expensive and time (and tissue) consuming, the medium and long-term benefits outweigh the initial expense and time.

Process Control Testing In Lieu Of Final Tissue Testing

True process validation testing can document the reliability of an improved disinfection or sterilization process that potentially allows for less relying on final tissue sterility testing. Using this approach provides a tissue bank with information on which of the various steps of the disinfection and sterilization steps of tissue processing most affect the final microbiological state of the tissue. Each critical step can be monitored or tested either instead of or with a reduction in routine testing of tissue at the end of the process. This monitoring or testing may range from microbiological analysis of a solution which has been in contact with the tissue to chemical analysis of a solution pre and post exposure to the tissue.

Thorough process validation can demonstrate that process monitoring is actually a better representation of the microbiological state of finished tissue because a greater quantity of samples can be tested. For example, microbiological testing of three different solutions which come into contact with the tissue during processing may represent the microbiological state of 100% of the finished tissue rather than only 10% of the tissue batch or lot, as is the case if 10% of the tissue is tested post-processing. In taking this approach one must address certain questions such as the sensitivity of a rinse method rather than an immersion method, but these items can be accounted for.

This approach may not eliminate all testing of finished tissue, but it can reduce the overall quantity of tissue which may be destroyed in testing. Routine testing of tissue at the end of processing demonstrates an uncertainty of the process to reliably provide microbiologically safe tissue. It demonstrates a lack of understanding of the capability of the validated process. With a careful understanding of the validated tissue processing steps from a scientific standpoint, contamination of the incoming allograft can be reliably eliminated. In a truly validated process, the safety which the process provides is well understood. However, a validated sterilization process can only be used if there is a thorough understanding of both the number and type of microorganisms present on the incoming tissue prior to the sterilization step, both at the time of initial validation and during the life of the use of the process as assured by regular monitoring of the incoming bioburden levels.

Microbial Testing of the Tissue Allograft Immediately Prior to Implantation

Some physicians have recommended culturing the tissue allograft immediately prior to implantation because of the worry that the allograft could be contaminated. Most tissue banks and tissue bank organizations do not recommend this step.

When the sampling is performed immediately before use, it takes days for the results to be available and the test result has limited clinical use. Rarely, it may aid in determining whether the allograft could have been the source of a postoperative infection. Aho et al. [13] reported two deep bacterial infections during use of 63 frozen unprocessed large bone allografts. *Pseudomonas aeruginosa* and *S. epidermidis* were isolated from the allograft immediately prior to surgery and later from the recipient site of infection. Both allografts had negative cultures at the time of recovery from the donor and prior to frozen storage. Perhaps these negative results were from inadequate sampling.

More often, a positive result from pre-implantation sampling can be a contaminant acquired from the operating room environment during sampling. A positive result can also be of limited or no importance when the isolate is a low-virulence bacteria or only a small quantity of bacteria is found and recipients who routinely receive prophylactic antibiotics and antibiotic irrigation solutions intraoperatively do not develop infections. Hou et al. [53] demonstrated that preimplantation testing can give positive results for bacteria that did not cause infections in patients. With the routine performance of pre-implantation cultures of thawed frozen bone allografts, they found that 22(1.6%) of 1,353 implanted allografts had a positive swab culture. Only four of these 22 patients (18.2%) developed infection. However, the bacteria found in wound cultures of the infected recipients were different from those found by the swab culture of thawed allografts except in one case. In this case the wound culture grew Candida and the allograft swab grew a yeast-like organism. Further testing was not performed to determine whether these coincidental findings of yeast occurred by chance or represented a contaminated allograft.

Preimplantation testing can give falsely-positive results creating serious but unneeded concern for patient safety and unneeded investigations and additional antibiotics given to the patient. In one interesting study by Mermel et al. [64], *Comomonas acidovorans* was found on four bone allografts when sampled immediately prior to implantation in four patients. Later, it was discovered that reports of bacterial growth were falsely positive and due to contamination of a water bath sonicator used in the microbiology laboratory to prepare tissue samples and not actually a contamination of the graft itself.

Summary

Microbial contamination of tissue allografts is a real risk and the use of microbial testing for bacteria and fungi by tissue banks is important to contain this risk. Microbial surveillance and monitoring includes testing of recovered tissues before processing, testing of the processing room environment, and final testing of the tissues. Careful validation of microbial reduction during each of the tissue processing steps can ensure safer tissue allograft use.

References

1. Eastlund T, Strong DM (2003) Infectious disease transmission through tissue transplantation. In: Phillips GO (ed) Advances in tissue banking, vol 7. World Scientific Publishing Company, Singapore. pp 51–131

2. Eastlund T (2005) Viral infections transmitted through tissue transplantation. In: Kennedy JF, Phillips GO, Williams PA (eds) Sterilisation of tissues using ionising radiations. Woodhead Publishing Limited, Cambridge, pp 255–278
3. Eastlund T (2006) Bacterial infection transmitted by human tissue allograft transplantation. Cell Tissue Bank 7:147–166
4. Cartwright EJ, Prabhu RM, Zinderman CE, Schobert WE, Jensen B, Noble-Wang J, Church K, Welsh C, Kuehnert M, Burke TL Srinivasan A (2010) Transmission of *Elizabethkingia meningoseptica* (formerly known as *Chryseobacterium meningosepticum*) to two tissue allograft recipients. J Bone Joint Surg (accepted for publication)
5. Centers for Disease Control and Prevention (2003). Invasive *Streptococcus pyogenes* after allograft implantation--Colorado, 2003. MMWR 52:1174–1176
6. James JIP (1953) Tuberculosis transmitted by banked bone. J Bone Joint Surg 35B:578
7. Kuehnert MJ, Clark E, Lockheart SR, Soll DR, Jarvis WK (1998) *Candida albicans* endocarditis associated with a contaminated aortic valve allograft: implications for regulation of allograft processing. Clin Infect Dis 27:688–691
8. Schotveld JH, Raijmakers AJ, Henry Y, Zaal MJ (2005) Donor-to-host transmitted Candida endophthalmitis after penetrating keratoplasty. Cornea 24:887–889
9. Al-Assiri A, Al-Jastaneiah S, Al-Khalaf A et al (2006) Late-onset donor-to-host transmission of *Candida glabrata* following corneal transplantation. Cornea 25:123–125
10 Harshvinderjit SB, Weinberg DV, Feder RS, Noskin GA (2007) Postoperative vancomycin-resistant *Enterococcuc faecium* endophthalmitis. Arch Ophthalmol 125:1292–1293
11. Centers for Disease Control and Prevention (2003) Clostridial endophthalmitis after cornea transplantation--Florida, 2003. MMWR 52:1176–1179
12. Tomford WW, ThongphasukK J, Mankin HJ, Feraro MJ (1990) Frozen musculoskeletal allografts. A study of the clinical incidence and causes of infection associated with their use. J Bone Joint Surg 72A:1137–1143
13. Aho AJ, Hirn M, Aro HT, Heikkila JT, Meurman O (1998) Bone bank service in Finland. Experience of bacteriologic, serologic and clinical results of the Turku Bone Bank 1972–1995. Acta Orthop Scand 69:559–565
14. Centers for Disease Control and Prevention (2001) Public health dispatch: update: unexplained deaths following knee surgery – Minnesota, 2001. MMWR 50:1080
15. Centers for Disease Control and Prevention (2002) Update: allograft-associated bacterial infections – United States, 2002. MMWR 5:207–210
16. Kainer MA, Linden JV, Whaley DN, Holmes HT, Jarvis WR, Jernigan DB, Archibald LK (2004) Clostridium infections associated with musculoskeletal-tissue allografts. N Engl J Med 350:2564–2571
17. Monafo WW, Tandon SN, Bradley RE, Condict C (1976) Bacterial contamination of skin used as a biological dressing. JAMA 235:1248–1249
18. Centers for Disease Control (2001) Septic arthritis following anterior cruciate ligament reconstruction using tendon allografts – Florida and Louisiana, 2000. MMWR 50:1081–1083
19. Centers for Disease Control and Prevention (1995) Ochrobactrum anthropi meningitis associated with cadaveric pericardial tissue processed with a contaminated solution – Utah, 1994. MMWR 45:671–673
20. Chang HJ, Christenson JC, Pavia AT, Bobrin BD, Bland LA, Carson LA, Arduino MJ et al (1996) *Ochrobactrum anthropi* meningitis in pediatric pericardial allograft transplant recipients. J Infect Dis 173:656–660
21. Christenson JC, Pavia AT, Seskin K, Brockmeyer D, Korgenski EK, Jenkins E, Pierce J, Daly JA (1997) Meningitis due to *Ochrobactrum anthropi*: an emerging nosocomial pathogen. A report of 3 cases. Ped Neurosurg 27:218–221
22. Reller LB, Maddoux GL, Eckman MR, Pappas G (1975) Bacterial endocarditis caused by *Oerskovia turbata*. Ann Int Med 83:664–666
23. McNeil MM, Brown JM, Carvalho ME, Hollis DG, Morey RE, Reller LB (2004) Molecular epidemiologic evaluation of endocarditis due to *Oerskovia turbata* and CDC Group A-3 associated with contaminated homograft valves. J Clin Microbiol 42:2495–2500

24. Pearson K, Dock N, Brubaker S (eds) (2008) Standards for tissue banking, 12th edn. American Association of Tissue Banks, McLean, VA
25. Martinez OV, Malinin TI, Valla PH, Flores A (1985) Postmortem bacteriology of cadaver tissue donors: an evaluation of blood cultures as an index of tissue sterility. Diagn Microbiol Infect Dis 3:193–200
26. McMahon CA, Lamberson HV (1989) Comparison of bacterial contamination of cadaveric bone donations collected under operating room and morgue conditions. In: Proceedings 13th Annual Meeting American Association Tissue Banks (October 1–4), Baltimore, MD, 1989
27. Bennett M, Johnson J, Novick S, Hilgren J, Rabe F, Eastlund T (1991) Prevalence and growth rate of microbes found at procurement of cadaver and living donor bone and connective tissue. Proceedings 15th annual meeting, American Association of Tissue Banks, Clearwater Beach, FL, 1991
28. Chapmen PG, Villar RN (1992) The bacteriology of bone allografts. J Bone Joint Surg 74-B:398–399
29. Scofield C, Klitzke K, Eastlund T, Steckler D (1994) Variables affecting bacteriologic contamination of tissue allografts acquired at procurement. Proceedings 18th annual meeting. American Association Tissue Banks, San Francisco, 1994
30. Vehmeyer SB, Bloem RM, Petit PLC (1994) Sensitivity and negative predictive value of swab cultures in musculoskeletal allograft procurement. Clin Orthop 300:259–263
31. Martinez OV, Malinin TI (1996) The effect of postmortem interval and manner of death on blood and bone marrow cultures from non-septic cadaver donors of tissues for transplantation. In: Proceedings of the 96th meeting American Society of Microbiology, New Orleans, LA, 1996
32. Deijkers RLM, Bloem RM, Petit PLC, Brand R, Vehmeyer SBW, Veen MR (1997) Contamination of bone allografts. Analysis of incidence and predisposing factors. J Bone Joint Surg 79-B:161–166
33. Bettin D, Harms C, Polster J, Niemeyer T (1998) High incidence of pathogenic microorganisms in bone allografts explanted in the morgue. Acta Orthop Scand 69:311–314
34. Journeaux SF, Johnson N, Bryce SL, Friedman SJ, Sommerville SM, Morgan DA (1999) Bacterial contamination rates during bone allograft retrieval. J Arthroplasty 14:677–681
35. Vehmeyer SBW, Bloem RM, Deijkers RLM, Veen MR, Petit PLC. (1999) A comparative study of blood and bone marrow cultures in cadaveric bone donation. J Hosp Infect 43:305–308
36. Vehmeyer SBW, Bloem RM (1999) Bacterial contamination of post-mortal bone allografts. In: Phillips GO, Kearney JM, Strong DM, Von Versen R, Nather A (eds) Advances in tissue banking, vol 3. World Scientific, Singapore, pp 33–41
37. Forsell JH, Liesman J (2000) Analysis of potential causes of positive microbiological cultures in tissue donors. Cell Tissue Bank 1:111–115
38. Vehmeyer SB, Bloem RM, Petit PL (2001) Microbiological screening of post-mortem donors – two case reports. J Hosp Infect 47:193–197
39. Vehmeyer S, Wolkenfelt J, Deijkers R, Petit P, Brand R, Bloem R (2002) Bacterial contamination in post-mortem bone donors. Acta Orthop Scand 73:678–683
40. Martinez OV (2004) Microbiological screening of cadaver donors and tissues for transplantation. In: Phillips GO (ed) Advances in tissue banking, vol 7. World Scientific Publishing, Singapore, pp 143–155
41. Martinez OV, Buck BE, Hernandez M, Malinin T (2003) Blood and marrow cultures as indicators of bone contamination in cadaver donors. Clin Orthop Relat Res 409:317–324
42. Malinin TI, Buck BE, Temple HT, Martinez OV, Fox WP (2003) Incidence of clostridial contamination in donors' musculoskeletal tissue. J Bone Joint Surg Br 85:1051–1054
43. Ibrahim T, Stafford H, Esler CN, Power RA (2004) Cadaveric allograft microbiology. Internal Orthop 28:315–318
44. van Baare J, Vehmeijer S, Bloem R (2004) Bacterial contamination of bone allografts in the Netherlands. In: Phillips GO (ed) Advances in tissue banking, vol 7. World Scientific Publishing, Singapore, pp 133–141

45. Fukusima R, Gianotti L, Alexander JW, Pyles I (1992) The degree of bacterial translocation is a determinant factor for mortality after burn injury and is improved by prostaglandin analogs. Ann Surg 216:438–445
46. Steffen EK Berg RD (1983) Relationship between cecal population levels of indigenous bacteria and translocation to the mesenteric lymph nodes. Infect Immun 39:1252–1259
47. Mejima K, Deitch EA, Berg RD (1984) Bacterial translocation from the gastrointestinal tracts of rats receiving thermal injury. Infect Immunol 43:6–10
48. American Association of Tissue Banks. Guidance Document No. 2: Prevention of contamination and cross-contamination at recovery: practices and culture results. American Association of Tissue Banks, McLean, VA. www.aatb.org.
49. Johnson D, Anderson M, Nelson N (2002) Factors affecting procurement culture results. Proceedings of the 26th annual meeting of AATB, Boston, Aug 23–27 62: Abstract PR-10
50. Nather A, Vikram D (2007) Femoral head banking: NUH Tissue Bank experience. Orthopedics 30:308–312
51. Prather J, Eastlund T, Steckler D, Stanek M, Reisman R, Umphress M (1990) Causes of discard of femoral head allografts donated by living donors in a regional surgical bone bank. Proceedings 14th annual meeting. American Association of Tissue Banks, McLean, VA
52. Salmela PM, Hirn MY, Vuento RE (2002) The real contamination of femoral head allografts washed with pulse lavage. Acta Orthop Scand 73:317–320
53. Hou CH, Yang RS, Hou SM (2005) Hospital-based allogenic bone bank: 10-year experience. J Hosp Infect 59:41–45
54. Scofield C, Eastlund T, Larson N, Steckler D, Korent H (1993) Causes of discard of femoral head allografts donated by living donors in a regional surgical bone bank. Proceedings 17th annual meeting, American Association of Tissue Banks, Boston, 22–25 Oct, 1993
55. Saies AD, Davidson DC (1990) Femoral head allograft bone banking. Aust NZ J Surg 60:267–270
56. Sommerville SM, Johnson N, Bryce SL, Journeaux SF, Morgan DA (2000) Contamination of banked femoral head allograft: Incidence, bacteriology and donor follow up. Aust NZ J Surg 70:480–484
57. Coors LA, Koster LA, Matthijsen NMC, Bokhorst AG, Van Wijk MJ (2008) Factors that influence blood culture contamination in cadaveric musculosketetal tissue donors. Proceedings of the combined 17th annual meetings of the European Association of Tissue Banks and the British Association of Tissue Banks, Edinburgh, Scotland, p 92. 12–14 Nov 2008
58. International Organization for Standardization. ISO 14644-1:1999 Cleanrooms and associated controlled environments – Part 1: Classification of air cleanliness
59. Farrington M, Matthews I, Foreman J, Caffrey E (1996) Bone graft contamination from a water de-ionizer during processing in a bone bank. J Hosp Infect 32:61–64
60. Ronholdt CJ, Bogdansky S (2005) Determination of microbial bioburden levels of pre-processing allograft tissues. In: Kennedy JF, Phillips GO, Williams PA (eds) Sterilisation of tissues using ionising radiations. Woodhead Publishing Limited, Cambridge, pp 311–318
61. International Organization for Standardization. ISO 11737-1:2006, Sterilization of medical devices – Microbiological methods – Part 1: Estimation of the population of microorganisms on product
62. United States Pharmacopeia. [71] Sterility Tests. In: USP/NF. United States Pharmacopeial Convention, Rockville, MD. 2003:2011
63. United States Pharmacopeia. [71] Sterility Tests. In: USP/NFp. United States Pharmacopeial Convention, Rockville, MD. 2003:2013
64. Mermel LA, Josephson SL, Giorgio C (1994) A pseudo-epidemic involving bone allografts. Infect Control Hosp Epidemiol 15:757–758

Part IV
Ensuring Quality by

Chapter 12
Establishing a Quality System

Scott A. Brubaker

Quality is not an act, it is a habit.

Aristotle (384 BC–322 BC)

At the core of every tissue establishment's philosophical and operational mantra must be the concept of "quality." Directed by an overarching quality management system and supported by the quality program that controls and assures compliance with established expectations, all employees of a tissue establishment should embrace the essentials of a working culture that promotes quality. Tissue banking professionals should understand that they must be advocates of quality because their establishment invariably offers multiple services and has customers and stakeholders at many levels; from donors and donor families to tissue banking coworkers and colleagues, as well as suppliers of equipment and reagents, and transplant professionals and tissue recipients. A tissue establishment today can provide services that are focused and limited (i.e., tissue recovery) or it can be a very complex operation that performs work encompassing an array of functions (i.e., obtaining donation consent/authorization; donor screening; infectious disease testing; tissue recovery; tissue processing and testing, storage and distribution) and this may involve various types of cells and/or tissue. Whatever donation role to which the establishment decides it will commit, it must apply quality concepts to its operations because, in countries where tissue banking is well developed, regulations have evolved and require it. It should be understood, however, that the successful application of the concept surrounding quality reaches beyond what regulations require. This chapter offers the author's perception of how this desire for the infusion of quality concepts in tissue banking has evolved and why it now pervades the cell and tissue banking profession. Additionally, suggestions are given that outline how a tissue bank's leadership should embrace "quality" and support the establishment's operations to ensure exemplary tissue practice is developed, performed, and maintained. In conclusion, lists demonstrate how continuous evaluation of your processes can provide

S.A. Brubaker (✉)
American Association of Tissue Banks, McLean, VA 22101, USA
e-mail: brubakers@aatb.org

G. Galea (ed.), *Essentials of Tissue Banking*, DOI 10.1007/978-90-481-9142-0_12,

assurance your activities are reliable and under control, and how systems should be developed to investigate failures and identify gaps leading to resolve and taking action to prevent recurrence. In tissue banking, exemplary practice is reached by a metamorphosis that blends scientific knowledge with an assiduous attitude for constant improvement. In essence, this provides a possible map for organizing your tissue establishment and providing quality tissue banking services.

Quality Tools

Two icons of the quality movement must be mentioned because their teachings form the basis that instill quality concepts into company management and the products and services organizations provide. In the 1950s, Joseph Juran described how quality is important to a firm's success. He later described a concept, called "The Juran Trilogy®" [1] (see Table 12.1), which combined quality planning, quality control, and quality improvement to form the foundation for managing successful operations. Also during this time, W. Edwards Deming described management methods that were similar but he simplified continual improvement using easily embraced actions: plan, do, check, and act. Both scholars described how this is an ever evolving process of managing while collecting, then analyzing, relevant information or data that feeds improvement to both your products and productivity, including cost control. Deming described a "System of Profound Knowledge" [2] (see Table 12.2) that management should follow for long term success and he described "14 points" [3], related principles that can support an effective business model.

Another teacher of total quality management and culture is John A. Woods and, in 1998, he published "The Six Values of a Quality Culture" [4] (see Table 12.3) in the fifth edition of the *Quality Yearbook*. He described that any business is a system

Table 12.1 Dr. Joseph Juran – "The Juran Trilogy®" [1]

Quality planning
• Determine who the customers are
• Determine the needs of the customers
• Develop product features that respond to customer's needs
• Develop processes that are able to produce those product features
Quality control
• Evaluate actual Quality Performance
• Compare actual performance to quality goals
• Act on the differences
Quality improvement
• Establish the infrastructure needed to secure annual quality improvement
• Identify specific needs for improvement – the improvement projects
• Provide the resources, motivation, and training to teams

Table 12.2 Dr. W. Edwards Deming – "System of Profound Knowledge" [2]

These interact:

Appreciation of a system

- Understanding the overall processes involving suppliers, producers, and customers of goods and services

Knowledge of variation

- The range and causes of variation in quality and the use of statistical sampling in measurements

Theory of knowledge

- The concepts explaining knowledge and the limits of what can be known

Knowledge of psychology

- Understanding concepts of human nature

Table 12.3 John A. Woods – "The Six Values of a Quality Culture" [4]

Value 1:	We're all in this together: company, suppliers, customers –versus "Everybody for him or herself"
Value 2:	No subordinates or superiors allowed –versus "The boss knows best"
Value 3:	Open, honest communication is vital –versus "Keep communication limited and secretive"
Value 4:	Everyone has access to all information on all operations –versus "Keep most information at the top, and share it only when necessary"
Value 5:	Focus on processes –versus "Focus on individual work"
Value 6:	There are no successes or failures, just learning experiences –versus "Success is everything, no tolerance for failure"

made of specific parts that interact and affect one another, and that the whole system must be viewed to best understand how to manage intelligently by promoting quality into the beliefs, values, attitudes, and behavior of employees and other customers. To change behavior, a realistic approach should be considered. Each value supporting a culture of quality is a logical extension of other values and the operations must be led by management that sincerely supports quality practice throughout the organization.

While all six values seem idealistic, they are values that take advantage of what each person in an organization has to offer and in understanding its structure as an interdependent system. These values embrace attributes that work together and lead to high quality operations: teamwork, loyalty, responsibility, personal identification with company, empathy with listening, increased knowledge and understanding, evaluation of process performance and improvement, and a positive attitude towards learning from experiences. Successful communication with internal and external customers that leads to process improvement is key to providing services that support your mission and goals. Understanding stakeholder and customer needs is essential for establishing the building blocks of your quality program. These quality concepts and others form the foundation from which standards and regulations throughout many professions are now based, including cell and tissue banking. Specifically, there is an expectation that we offer professional services that fulfill the public's wish to donate and this extends through our activities to the proviso that we distribute safe and effective cells or tissue allografts for transplantation – these hopes (the mission) can be met by effectively applying quality tools to your tissue establishment's operations like those described above. This is an expectation of sound quality management.

A Sequential, Historical Perspective – Standards and Regulations

Standards are principles that are developed, agreed upon, and published that delineate values and expectations to achieve best practice. This is akin to reaching established goals supported by good reasoning; a cookbook for attaining excellence...or, high quality. For cell and tissue banking, the development of the first edition of AATB's *Standards for Tissue Banking* [5] *(Standards)* in 1984 established the first formal publication of such best practices for our profession. This first edition of published standards was borne from a fundamental element of any quality program; written procedures. Provisional guidelines for proposed standards were developed between 1978 and 1981 and these procedures described tissue banking operations as applied to various tissue types. Although these *Standards* were published, a "procedures manual" was also issued (1986) by the AATB that continued to offer standardized methods for handling tissue. Soon, Technical Manuals (1990–1992) were organized and released that described step-by-step procedures to successfully accomplish tissue banking operations with an emphasis on processing methods and donor screening. Eventually, elements of the Technical Manuals were absorbed into the *Standards* and by 1996, the seventh edition, an evolution occurred that reflected a new requirement to include "quality assurance and quality control programs" and to apply this concept to all tissue banking functions. New headings in *Standards* included: Records Management; Release and Transfer of Tissue; General Operations (that included requirements for a procedure manual, staff training and competency; safety practices; and for facilities and equipment). These additions mandated a quality systems approach and resembled concepts like those found in "good manufacturing practice," or GMP, which had become required of medical device and pharmaceutical manufactures in recent years. A reason this transformation occurred to these voluntary *Standards* was because a handful of AATB-accredited tissue banks had each been designated by the United States (US) Food and Drug Administration (FDA) as a "manufacturer of a replacement heart valve." In the early 1990s, cryopreserved allograft heart valves had suddenly become classified by the agency as Class III Medical Devices, the strictest device classification. These tissue banks experienced the requirement of implementing

GMP into tissue banking functions so it was logical to incorporate these concepts into the *Standards* for all cell and tissue types, regardless of classification (or no classification). Additionally, members of the AATB who were in key positions (the President was Ted Eastlund, M.D., and the Chairperson of the Standards Committee was Jeanne V. Linden, M.D., M.P.H.) had crossover involvement as blood banking professionals. This is important because during this same time in the early 1990s, blood establishments in the US began to experience an expanding federal oversight, described in the FDA's "Blood Action Plan," [6] that applied quality systems to blood banking operations; these are similar to those found in the pharmaceutical industry. In response to these changes and to better serve their membership, the AABB (formerly known as the American Association of Blood Banks) identified ten Quality System Essentials (QSEs) as the minimum elements that must be addressed in a quality system and began structuring the organization of its standards based on these QSEs. This adjustment was evident with publication of the 18th edition in 1997 of AABB's *Standards for Blood Banks and Transfusion Services*. Blood donation and the services offered that ultimately lead to safe transfusions for millions of recipients bear many similarities to cell and tissue donation, banking, and transplantation. The US FDA soon followed with the publication of a similar "Tissue Action Plan" [7] in 1998 that described the eventual development of "good tissue practice" which is described later as reaching publication in 2004.

Prior to the issuance of a majority of publications that addressed quality systems for cell and tissue banking, other professions had a history of applying these concepts. As tissue banking professionals, we can apply knowledge gained from the application of quality systems to the operations of these professions (medical device, pharmaceutical). Because medical devices, pharmaceuticals, blood products, and cell and tissue allografts all have human recipients in common, the importance of controlling the safety and quality of these products is paramount. A recipient adverse reaction could result in morbidity or mortality so it makes sense to require quality systems that should be built to avoid disasters. Prior to applying these quality concepts to cell and tissue banking, publications aimed at manufacturers existed including the voluntary standards issued by ISO, the International Organization for Standardization. The ISO 9000 series of standards [8] for quality assurance and quality management were first published in 1987 and had a dramatic international impact due to their scope and rate of acceptance. This series of standards focuses on processes, procedures, and practices and is applicable to all sizes and types of manufacturing and service companies. Eventually, the ISO 9000 family of standards was approved by CEN, the Comité Européen de Normalization, as a European Standard covering quality management systems. Its current, main components are listed in Table 12.4.

In general, quality management systems (QMS) should: (1) provide compliance with regulatory requirements, (2) demonstrate a serious commitment to quality, (3) provide safe products, and, (4) services and products should satisfy customers. A QMS is comprised of (1) proper organizational structure; (2) delineated responsibilities; (3) written procedures and processes; and, (4) necessary resources. Management must be able to listen to stakeholders, keep them well informed so they

Table 12.4 ISO 9000 quality management systems (QMS) family of standards [8]

ISO 9000:2005 QMS – Fundamentals and vocabulary

- Explains advantages of implementation, increased confidence and satisfaction, the mutual understanding of terminology, requirements to audit for conformity, and guidelines for standards development and personnel training

ISO 9001:2008 QMS – Requirements

- Describes management responsibilities, product and service realization, and use of measurement and analysis, all of which lead to continual improvement and enhanced customer satisfaction (this is the standard to which ISO registration is certified)

ISO 9004:2000 QMS – Guidelines for performance improvements*

- Provides guidance and recommendations that focus on the achievement of ongoing improvement, measured through the satisfaction of customers and other interested parties (Note: the work plan by TC 176, the technical committee maintaining ISO 9004, includes a plan to update this standard by August 2009).

can realize when improvements should be made, and personnel must be empowered to make those improvements. A downfall of some organizations is that their departments, services, and/or suppliers work in "operational silos." This isolation can be non-productive and be a cause for errors, goals not being met, and dissatisfied stakeholders. The concept is simple and applicable to cell and tissue banking operations but successful application can be complex. There are many tissue bank operations that are connected that include a variety of people performing specific functions along the pathway that makes successful donation and transplantation possible. If these folks work in functional silos, failures will occur that are most likely avoidable. If professionals who perform sequential functions only communicate with whom they have direct contact, the entire system may not be effective and anticipated goals may not be realized. For instance, if donation laws are not well understood or education about them is lacking, the public may not realize their options and this can affect the end user clinician who will not have access to a sufficient amount of cells or tissue allografts needed for recipients. Who is responsible for communicating this information to the public? Is it only those persons who have direct access to potential donors or is it related to a wide variety of healthcare professionals who handle donated cells and tissues (e.g., cell/tissue processors, health care management professionals, surgeons that use allografts)? Is development of a referral process only limited to the front line people who would look for possible donors or should other stakeholders be involved and be visibly supportive? How are needs of end users (surgeons) or outcomes of recipients known to the tissue bank's processing technicians who decide what form the allograft will take? Tissue bank professionals who recover and send tissue to another tissue establishment for processing should not only understand the fundamentals of their functions but they should also be informed regarding how the tissue can be processed, how it can be used, and be informed of specific successful applications. Those who recover tissues should be educated in regard to the expectations of colleagues who receive the tissue (e.g., control of contamination and cross-contamination, proper tissue wrapping and labeling, technical specifications, use of proper reagents and materials, maintaining proper environmental transport temperatures, etc.). Suppliers should be aware that if they change specifications for equipment or reagents they provide to a tissue establishment, or if they substitute an item for the usual one ordered, they must be certain these changes are acceptable for their client (the tissue establishment). But, how can this be controlled? Providing written agreements, contracts, and/or standard operating procedures that delineate responsibilities and expectations are helpful, but periodic auditing is essential to monitor compliance. These links between performance of functions and the extensive intercommunication that should occur to enhance the overall understanding of stakeholder expectations, includes everything from donation to recovery and processing, through storage and use or other final disposition of the cells or tissue. A tissue establishment's well-developed QMS can identify and prevent these gaps and improve customer satisfaction both internally and externally.

Other ISO standards describe QMS and can be helpful when applying a quality system approach to tissue banking operations. One such standard is *ISO 13485:2003 Medical Devices – Quality Management Systems – Requirements for Regulatory Services* that superseded EN 46001 and EN 46002 published in 1997 and ISO 13488 from 1996. The EN (European Norm) standards are only applicable to the European Union but were very much liked by ISO so they were replicated when developing this ISO standard and this harmonized them to an internationally recognized standard. These two EN standards are equivalent to ISO 9001 and 9002, except they are specific to manufacturers of medical devices. This basic content also made its way into the GMP regulations for medical devices of the US, Japan and Canada. Quality system requirements mimic those of its related ISO 9000 family of standards.

At about this same time, in 1996, the US FDA introduced Quality System Regulation (QSR) for Medical Devices (§1271.820) with related requirements at §1271.803 Medical Device Reporting, §1271.806 Medical Device Reporting Corrections and Removals, and §1271.821 Medical Device Tracking. After recognizing specific needs, the FDA published: the Medical Device Quality Systems Manual – A Small Entity Compliance Guide; Guidelines on Principles of Process Validation; the FDA Worldwide Quality Systems Requirements Guidebook for Medical Devices; and the Quality Systems Inspection Technique (QSIT). By 1998, Canada issued similar medical device regulation but elected to refer to quality management and quality assurance standards such as ISO and equivalent standards of the Canadian Standards Association (CAN/CSA-ISO 13485, and 13488). Similar directions and requirements, all related to QSM standards adopted internationally, eventually appeared in various global human tissue and cell regulations within the next 10 years.

In 1997, the Treaty of Amsterdam mandated in Article 152 at (4)(a) that "The council, … ,shall contribute to the achievement of the directives referred to in this Article through adopting: Measures setting high standards of quality and safety of organs and substances of human origin, blood and blood derivatives; these measures shall not prevent any Member State from maintaining or introducing more stringent protective measures." As a "substance of human origin" human tissues was included in the plan to develop regulations for the communities of the European Union that would be shaped by the European Parliament. This eventually led to development and publication of the mother Directive for human tissues and cells, "*Directive 2004/23/EC of the European Parliament and of the Council of 31 March 2004 on setting standards of quality and safety for the donation, procurement, testing, processing, preservation, storage and distribution of human tissues and cells*" [9]. The word "quality" appears thirty times in this Directive, and "quality and safety" appear together twenty-four times. Quality Management is described in Article 16 and, at a minimum, includes the following documentation for the quality system: (1) standard operating procedures, (2) guidelines, (3) training and reference manuals, (4) reporting forms, (5) donor records, and (6) information on the final destination of tissues or cells. Before this Directive was developed, there were mandates issued in various countries that provided guidelines to promote quality for tissue banking operations.

Since there are global regulations that designate products containing cells and tissues as medical devices, it's natural that tissue banks implement GMP-like practices into daily operations. These are being developed and are called GTP (good tissue practice). By way of US FDA tissue establishment registration regulations published in 2001, if a cell or tissue graft is combined with another regulated product for its application, or use is determined to be non-homologous [10], the combination graft may be determined to be a biologic and must follow applicable GMP, or it can be determined to be a medical device and, additionally, the medical device regulations, including QSR, must be followed. These "product" determinations can classify a tissue bank as a medical device manufacturer. This additionally supports the rationale to institute a quality program concept into conventional tissue banking activities, previously incorporated by AATB's *Standards* in the mid 1990s. I should comment that tissue and cell banking professionals are generally not amenable to referring to allografts as "products" or "devices" and we don't particularly care to be described as "manufacturers." Unfortunately, regulatory jargon that must be used trumps our inherent application of an emotional, human dimension to our work. To use regulatory language demands that we have to adjust, but this doesn't mean we have lost sight of our morals or respect for humanity. The quality culture instilled in each tissue/cell bank employee should include this respect and honor of donors, donation, and future recipients.

The Australian Code of Good Manufacturing Practice (for) Human Blood and Tissues [11] are regulations published in 2000 by the Therapeutic Goods Administration (TGA) and contains GMP expectations for tissue banks. The requirements for a Quality System are described and include: quality objectives, organizational structure, monitoring systems, and management review. The Code includes the elements of quality systems from the ISO 9000 series of standards and applies these principles to blood and tissue banking. Similar to designations made of human cell and tissue products by the US FDA, allografts can be regulated as medicines or therapeutic devices, depending on their biological and mechanical properties, or their therapeutic purpose. Further regulations for human cell and tissue banking are planned for development and publication by the TGA [12].

The next regulations to inject quality systems requirements into tissue banking activities were issued by the United Kingdom's Department of Health with publication in 2001 of *A Code of Practice for Tissue Banks (providing tissues of human origin for therapeutic purposes)* [13]. It covered requirements for: (1) Quality Systems; (2) Tissue Bank Facilities; (3) Responsibilities of Personnel and Training; (4) Donor Selection; (5) Control of Tissue, Services, and Materials; (6) Process Control; (7) Packaging, Labeling, and Transport; (8) Documentation; and, (9) Special Considerations.

Human tissue allografts can also be classified "medicinal products" as was adopted in recent years by regulatory authorities in Germany and Austria. This designation surpasses requirements of the European Union (EU) human tissue and cell Directives to be described later. In Commission Directive 2003/94/EC [14], the rules governing medicinal products for human use in the EU are described and require conformity with good manufacturing practice. Definitions for "medicinal product" and "substance" can be interpreted as applicable to human tissues and cells (see reference to Directive 2001/83/EC [15], and consider all revisions), but specific references to human "tissue," "cells," or "transplant" is not made. One of three Directives eventually published for human tissues and cells for transplantation (Directive 2006/86/EC) references 2003/94/EC for specifications regarding air quality standards in the processing area. For further guidance when interpreting GMP requirements in 2003/94/EC, see Volume 4, *EU Guidelines to Good Manufacturing Practice, Medicinal Products for Human and Veterinary Use* (see revisions in 2005 and 2008) [16]. Chapter 1 describes requirements for Quality Management and an addition in early 2008, Annex 20, describes Quality Risk Management (QRM) as required within the quality system framework. This annex corresponds to ICH Q9 guidelines for QRM that provide principles and a framework for decision-making regarding perceived "risk." It's a useful quality improvement methodology and it supports science-based decision making when evaluating risk. ICH is the International Conference on Harmonization of Technical Requirements for Registration of Pharmaceuticals for Human Use. It aims to maintain safeguards on quality, safety, and efficacy, and regulatory obligations that protect public health on a global scale. By applying this to tissue banking, QSM can be realized and offers substantial controls. Persons responsible for Quality Management of a tissue or cell bank should be informed and apply these concepts.

It must be described that the establishment of voluntary standards, and implementing them into your tissue banking operations via policies and procedures, adds much value to your quality system. As national regulations and other documents have been produced that govern cell and tissue banking, regulators have relied on the collective expertise of tissue banking professionals and have referenced published "industry" standards that guided them in developing regulation. This has been most evident in the US, Europe, and Canada. Professional standards and guidelines have been produced as long ago as the mid 1980s (AATB) but other tissue and cell banking groups have also issued standards that have guided their constituencies by providing guidelines and promoting safe practice and this has preceded government regulations. This resulted in a system of professional self-regulation for over a dozen years. From the mid 1990s through 2002, the following cell and tissue banking professional associations have published standards: EATB (European Association of Tissue Banks); BATB (British Association for Tissue Banking); AEBT (Asociación Española de Bancos de Tejidos, aka Spanish Association of Tissue Banks), EAMST (European Association of Musculoskeletal Tissues); EEBA (European Eye Bank Association); JACIE (Joint Accreditation Committee – ISCT and EBMT, these are the International Society for Cellular Therapy and the European Group for Blood and Marrow Transplantation); as well as one of the first standards-setting groups, the EBAA (Eye Bank Association of America). The Council of Europe also began to publish guidelines via "Recommendations and Resolution" documents in 1994 that led to the first edition in 2003 of the *Guide to Safety and Quality Assurance for Transplantation of Organs, Tissues, and Cells* [17]. This guide is now in its 3rd edition (2007) and thoroughly describes the need for a QMS. To assist all countries where cell and tissue banking activities are developing and that might be without regulatory oversight for it, the WHO (World Health Organization) convened meetings in 2004 and 2005 of tissue banking experts from numerous countries. They gathered to discuss global standards necessary for the development of safe tissues and equitable and ethical access to donation and transplantation of cells and tissues. These meetings produced two useful "Aide Memoires" for use by national health authorities. In 2005 *Key Safety Requirements for Essential Minimally Processed Human Cells and Tissues for Transplantation* [18] was published and in 2006 *Access to Safe and Effective Cells and Tissues for Transplantation* [19] was made available. Among other key elements, both documents promote quality systems and quality programs. The importance to enhancing the safety of tissues and cells used for clinical applications that has been provided by these associations/organizations via publications of standards and guidelines cannot be stressed enough. On a global scale, cell and tissue banking professionals promoted safety and quality long before it was required by regulations.

Between 2004 and 2007, definitive, key cell and tissue regulations were published in the US, Europe, and Canada. In 2004, Subpart D of the US FDA's cell and tissue regulations (codified at 21 CFR 1271) was published and is a Final Rule describing Current Good Tissue Practice [20] (CGTP) requirements. It contains a description of "core CGTPs" that can directly control the introduction, transmission or spread of communicable disease and these include: Facilities; Environmental control and monitoring; Equipment; Supplies and reagents; Processing and process controls; Recovery; Processing and Process Controls; Labeling controls; Storage; Receipt, predistribution shipment, and distribution of an HCT/P; and, Donor eligibility determinations, donor screening, and donor testing. Additionally, it includes functions that support this control and these are titled: Establishment and maintenance of a quality program; Personnel; Records; Tracking; Complaint files; Procedures; Process changes; Process validation; and Exemptions and alternatives. FDA now basically required in regulation what AATB's *Standards* had implemented 9 years prior. It is interesting to note the differences between requirements found in CGTP versus those in Quality Systems Regulations (for products determined to be "medical devices") and similar requirements in Current Good Manufacturing Practice (for products determined to be "biologics"). A dozen such differences were listed as requirements found in CGTP that are not completely covered in QSR or CGMP. The lists are found in FDA's *Draft Guidance for Industry: Current Good Tissue Practice (CGTP) and Additional Requirements for Manufacturers of Human Cells, Tissues, and Cellular and Tissue-Based Products (HCT/Ps) (1/16/2009)* [21]. Among the list are these, which make sense since human tissues and cells are a very different source material than metals and plastics (manmade raw materials): all donor eligibility requirements; prevention of the introduction, transmission, or spread of communicable diseases; prohibition on pooling; predistribution shipment; HCT/P availability for distribution only after donor eligibility established; and record-keeping for 10 years (and facility cleaning and sanitation records for 3 years). A related document, *Guidance for Industry – Validation of Procedures for Processing of Human Tissues Intended for Transplantation* [22] was issued a few years earlier (2002) due to FDA's need to offer some instruction in response to tissue allograft recipient adverse reaction reports linked to processing deficiencies [23, 24]. Specifically identified was inadequate validation of microbiological culturing techniques [24, 25]. Although the FDA's Guidance lacked the instructional detail needed regarding *how* to perform adequate validations involving human cells and tissues, it did offer general expectations of FDA by describing that a tissue establishment can obtain data to document the effectiveness of a procedure being used to prevent contamination. These three examples are given:

Verifying full and proper implementation of a previously validated procedure such as may be found in a technical manual of another organization, or

Conducting literature searches to demonstrate that the procedures implemented are known to be effective in preventing the infectious disease contamination (e.g., Environmental Protection Agency-approved chemical sterilants for laboratory surfaces), or

Conducting off-line or on-line challenges with indicator organisms, as appropriate, or evaluating the capacity of the manufacturing process to prevent contamination during processing.

At this writing, the AATB has formed a task force that is developing guidance covering all aspects of a "microbiological surveillance program," which includes specific guidelines for process validation and validation of culture methodologies used for various human tissue types.

The landmark publication of Directive 2004/23/EC [9] in March of 2004 described, among other directions, requirements that a tissue establishment have a quality system supported by an oversight of quality management. The quality message is clear throughout the articles and there is a description that a "responsible person" must be qualified and has responsibility for: procurement; testing; processing; storage; obligations to the competent authority; and for implementing quality management as described in various articles that include oversight of personnel, inspections, procedures, regulations, third party agreements, technical requirements, and handling serious adverse events and serious adverse reactions. The definition of *quality system* is described in *Article 1* at (d) that it "means the organisational structure, defined responsibilities, procedures, processes, and resources for implementing quality management and includes all activities which contribute to quality, directly or indirectly." A definition for *quality management* appears in *Commission Directive 2006/86/EC of 24 October 2006* [26] and it "means the coordinated activities to direct and control an organisation with regard to quality." In this Directive, a description of requirements are also made available in *ANNEX I* that includes detailed directions on how to acquire and maintain a quality management system and it covers: organization and management; personnel; equipment and materials; facilities/premises; documentation and records; and, quality review. Important quality-related definitions appear in *Commission Directive 2006/17/EC of 8 February 2006* [27] as well, such as "validation" (or "qualification" in the case of equipment or environments) and "traceability," and it contains procedures as well as selection criteria and laboratory test requirements for donors. Another Directive was issued in late 2007 titled *Regulation (EC) No 1394/2007 of the European Parliament and of the Council of 13 November 2007 on advanced therapy medicinal products and amending Directive 2001/83/EC and Regulation (EC) No 726/2004* [28]. This regulation supplements the requirements of the mother and two daughter Directives listed immediately above and is applicable to "advanced therapy medical products." Included in that definition are gene therapy products, somatic cell therapy products, and tissue engineered products. It describes classifications for combination products, lays down further controls (labeling, post-authorisation efficacy) and requires definitive regulatory review (evaluation of characteristics) for such products since they are more than minimally manipulated and can have a non-homologous function when used. This is analogous to formal review and designation of such products that is required by the US FDA.

During these years of developing tissue and cell regulations in North America, throughout Europe, and in Australia, the International Atomic Energy Association (IAEA) was developing their *International Standards for Tissue Banks* [29] and it began to be referenced by tissue banking professionals in thirty countries throughout the regions of Latin America and Asia/Pacific, the Middle East, and in Africa. From the early versions of these *International Standards for Tissue Banks* through to final publication in 2005, it contained a description for the need to have a Quality Management System. Basic Elements of an Appropriate Quality Management System were listed as: Organizational Structure and Accountability; Documentation; Control of Processes (SOPs); and Record Keeping. These elements were described in great detail within the standards and there is direction given to consult the ISO 9000 series of standards for Quality Management and GMP. Quality Requirements (Quality Assurance, Quality Control Programme) were also referenced. The publication in 2007 of the IAEA *Radiation Sterilization of Tissue Allografts: Requirements for Validation and Routine Control – A Code of Practice* [30] is of extreme importance in tissue banking on a global scale. The entire document is dedicated to quality control recommendations for this technically challenging tissue treatment. Particulars regarding qualification, verification, and validation are well described. The incredulous training programs related to these documents that the IAEA funded and organized are truly some of the most successful to date. The curriculum was offered via internet as well as face-to-face courses with the possibility to acquire a university diploma. Today, the course is available on compact discs.

Continuing with development of national regulations, in June 2007, Health Canada published the *Safety of Human Cells, Tissues and Organs for Transplantation Regulations* [31] in the *Canada Gazette.* A draft Guidance was also released at this time and this was finalized in early April of 2009 as the *Guidance Document for Cell, Tissue and Organ Establishments, Safety of Human Cells, Tissues and Organs for Transplantation* [32]. The cell, tissue and organ (CTO) regulations include a definition for "quality assurance system," which is required. In 2000, Health Canada contracted with the Canadian Standards Association (CSA) to facilitate the publication of National Standards designed to maximize the safety, quality and performance of cells, tissues, and organs for transplantation. The Standards Council of Canada has accredited the CSA as the standards development organization in Canada, and CSA standards are referenced in parts of the CTO regulations. The standards in CAN/CSA *Z900.1-03* [33] describe the need for a Quality Assurance Program, Quality Control, and oversight responsibility by Quality Management. The need for standard operating procedures and control of records is emphasized. And, specifically for tissues, in CAN/CSA *Z900.2.2-03 Tissues for Transplantation* [34], responsibilities for final Quality Assurance/Quality Control review to release tissue is described in some detail.

Significant Quality Projects

For 10 years, the AATB has organized and held an annual Quality Assurance Workshop where quality issues specific to tissue banking have been presented and discussed by professionals primarily representing quality program and regulatory affairs departments at tissue banks. Practical, every-day topics (e.g., quality auditing, training, performance measures, complaint handling, SOPs, etc.) have been on the agenda with information sharing via open discussion and role-play that challenge individual perspectives. In recent years, this workshop has been co-sponsored by the EBAA since ocular tissue and conventional tissues (bone, soft tissue, skin, cardiac, vascular) encounter similar thought-provoking operational situations.

In Europe, the European Quality System for Tissue Banking (EQSTB) [35] project began in 2004 and can be summarized by the following four specific objectives it had:

1. Analyze standards or guides used in different European tissue banks and focus on finding similarities and differences among tissue banks' protocols. A recommendation of standard operating procedures will be prepared based on the European Directives for Human Tissues and Cells.

2. Build a Tissue Registry through a multinational European network database. Standardized data and protocols will be suggested to receive the same information from different tissue banks regarding donor selection, tissue retrieval, processing and transplantation.
3. To design and validate a specialised training model to be used for tissue bank personnel that can become the approved education recommended by European Union members. This model aims to be a method of qualifying personnel and, finally, certify their knowledge.
4. To create a pilot model for European Accreditation of tissue banks based on the concepts of European Directive 2002/0128 (COD) that establishes a comparable national inspection and accreditation structure.

This comprehensive program was developed by dedicated, knowledgeable tissue banking professionals representing many Member States and, after completing its 3-year course, was a success. It produced two essential guides:

The *Guide of Recommendations for Tissue Banking* [36] (2007) that covered:

- Quality systems
- Legal and regulatory framework
- Standards
- Quality and safety key points

And, the *Guide for Auditing Tissue* Establishments [37] (2007) that included tools for:

- Self-assessment
- Internal audits
- External peer audits
- Third party audits

Another important Commission funded project that remains active is named EUSTITE (European Union Standards and Training in the Inspection of Tissue Establishments [38]). The primary objective of the EUSTITE project is to optimise and harmonise the standards and methods applied by Competent Authorities in the inspection and accreditation of tissue procurement and tissue establishments within the EU, in compliance with Directive 2004/23/EC, Articles 5, 6 and 7 and its associated implementing directives. A secondary objective is to propose common systems for definition, classification and reporting of adverse events and reactions that are consistent with similar systems in other parts of the world. The project has four main components:

- Achieving consensus on best practice in the inspection of tissue establishments;
- The development of practical guidelines for the conduct of inspections in tissue establishments;

- The design and testing of a training programme for inspectors in the field;
- The establishment of a pilot scheme for adverse event and reaction reporting and management.

To date, this project has finalized publication of these very useful tools:

- The *Inspection of Tissue and Cell Procurement and Tissue Establishments – Guidelines for Competent Authorities (2008)* [39]. It includes sections addressing Qualification and Training of Inspectors, Inspection Scheduling, Type of Inspection, Conduct of Inspections, and Evaluation of the Inspection System, plus multiple descriptive Annexes.
- *Tools for Vigilance and Surveillance of Human Tissues and Cells (2008)* [40]. It addresses: Roles and responsibilities in the Management of Serious Adverse Events and Reactions in the European Union; Triggers for Reaction/Event Reporting; Communication with stakeholders; Tools that assist with Severity Grading, Imputability Assessment, and Impact Assessment for adverse events and reactions; Reporting Forms; Evaluation of Vigilance Systems; and provides Annexes.

The EUSTITE Project is unique and groundbreaking for the cell and tissue banking profession. Participants in our field should become familiar with the documents it has produced and incorporate these quality systems into the quality programs at tissue establishments everywhere.

Practical Applications of QMS and QC for Tissue Establishments

When reviewing regulations, standards, and guidance, the tissue establishment's "responsible person" is given the task to apply these concepts to her or his specific operations. This is not a small job and it is a big responsibility. Although one person may be designated as "responsible," the fact is that all employees must be responsible. The quality culture concepts [4] described earlier apply and instilling this culture is the responsibility of management. This is not to say that the establishment's Quality Management is not ultimately responsible, but this person (or group of managers) need(s) to thoroughly understand the interaction between the operations for which they are in charge, how quality can change due to a variety of influences, and be knowledgeable of inherent limitations. This is the concept of Deming's "System of Profound Knowledge [2]." Management is also expected to be able to develop, control, monitor, and improve quality systems that specifically relate to their firm's functions (see The Juran Trilogy® [1]). Listings and tables follow that offer examples of possible ways to apply QMS and QC to tissue and cell banking functions. These are not intended to be all encompassing and applicability can differ, but these lists can be used to evaluate systems.

General Quality Management Responsibilities

Can include: (1) *Active* oversight of quality system/quality program; (2) Fiscal responsibility, and (3) Future planning. Components of these responsibilities may include:

(1) *Active* oversight of quality system/quality program (commitment to quality)
   - Organizational chart
     - Defined responsibilities for Quality System Management
     - Personnel: qualified, experienced, skilled, trained, competent
   - Establish quality policy and quality objectives
     - Measurable objectives that meet policy; periodic review
       - Suitability and effectiveness of QP (monitoring system)
     - Develop and maintain Quality Manual
   - Ensure "quality" communication to and within the organization
     - Define responsibilities and authority
     - To appropriate levels
     - Stress importance of the customer, regulatory, and legal requirements
   - Data collection and dissemination/sharing
     - Audits, various measurements/monitoring, data analysis, complaints, recalls
       - Includes handling of serious adverse events/reactions (SAE, SAR)
     - Continuous Quality Improvement and initiatives (e.g., CAPA)

(2) Fiscal responsibility (budget)
   - Must show evidence of commitment to ensure proper resources
     - Infrastructure/personnel
     - Facilities, equipment, supplies; sufficient number of personnel; etc.
   - Contracts
     - Define expectations and responsibilities, include a detailed Quality Agreement

(3) Future (strategic) planning
   - Annual re-assessment of mission
   - Organizational viability
     - 5-year plan
     - Growth and development
     - Plan in case of "disaster" (internal or external influences)

Considerations for a Tissue Establishment's QMS

- The Quality Program includes: the Quality Manual; regulatory compliance; master validation and qualification plan; audit plan (internal and external); computer/software verification/validation; identification of critical functions and establishment of performance indicators for all critical operations; CAPA program; traceability of tissue from donation through final disposition; and handling recalls
- Facility(ies) management includes: adequate size; properly equipped; good location; maintained (cleaned, repaired); facility inspections and schedules; and performance of environmental monitoring, where applicable, at designated intervals
- Personnel management: establishing qualifications; induction and ongoing training/re-training; training manuals; competency documentation via performance audits/written tests; designation of authorized persons for critical functions; thorough position descriptions; and, provision of an adequate number of staff
- Equipment, supplies, and reagents: establishing specifications; qualification (equipment IQ, PQ, OQ* when indicated); maintenance (calibration, preventive maintenance); supplier qualification; and systems management, such as: water, steam, ventilation, engineering, irradiation
- Policies and Procedures: includes work instructions and availability to staff; document control; change control; data handling (cover all relevant functions where possible)
- Third-party agreements (written contracts or other arrangements): responsibilities and expectations of all parties must be clearly delineated; periodic quality audits of third parties; and, inclusion of a "quality agreement" is desirable in every contract

Examples of Third-Party Agreements (Written Contracts or Other Arrangements)

Suppliers

– Controlled selection process (supplier qualification checklist); set supply specifications

Services

– Donor referral
– Donor screening
– Consent/authorization
– Tissue recovery
– Tissue processing
– "Testing" (laboratories)- Donor: infectious disease; Tissue: cultures, quality measurements
– Instrument sterilization/tissue irradiation services
– Equipment setup, calibration, and maintenance: IQ, OQ, PQ* (refrigerators, freezers, alarms, sterilizers, etc.)
– Biohazardous waste disposal
– Facility cleaning
– Transportation (tissue, or personnel)
– Distribution: import or export of tissue; recalls; intermediary establishments

*IQ, OQ, PQ = Installation Qualification, Operational Qualification, Performance Qualification

Quality Control (**QC**) – Specific tests defined by the QA Program to be performed to monitor recovery, processing, preservation and storage, tissue quality, and test accuracy. These may include but are not limited to, performance evaluations, inspection, testing, and controls used to determine the accuracy and reliability of the tissue bank's equipment and operational procedures, as well as the monitoring of supplies, reagents, equipment, and facilities.

From A2.000 Definitions of Terms, AATB *Standards for Tissue Banking*, 12th edition, 2008

Tissue Establishment Functions or Practices That Could Be QC'ed (*Written Procedures Required for All Functions*)

- Donor referral
- Consent/authorization
- Donor screening/evaluation (includes physical assessment)
- Donor testing– qualification of blood sample (check for plasma dilution), required tests, approved test kits, qualified testing laboratory, are results shared as appropriate?
- Tissue recovery (procurement)/wrapping/labeling/transport
- Tissue receipt and acceptance/storage/quarantine
- Tissue processing/preservation/packaging/labeling/storage (monitoring)
- Tissue testing: culture results; quality results such as residual levels (moisture, calcium, potency claims, etc)
- Facility controls – Environmental control or monitoring, etc
- Equipment maintenance including IQ, PQ, OQ*
- Quality Assurance Review – donor and tissue record reviews

  Release of tissue from quarantine
  Proper destruction of non-conforming tissue
- Distribution and transport of tissue
- Receipt of utilization/final disposition information
- Tracking donor and tissue from start to final disposition (unique ID/coding)

Examples of Tissue Establishment QC Programs

- Standard Operating Procedures (SOPs)
  - Contain instructions for steps, materials, and methods
    - Use a standardized format
      - Title, ID/numbering, signatures/reviews, goal, definitions, etc.
      - Includes safety precautions, when applicable
      - Describes relationship to other polices, procedures or work instructions, when applicable
  - Must be readily available to personnel performing the functions/steps
  - Must be validated, when applicable (i.e., related to processing)
      - See "master validation plan"
  - May only need "full verification" that required materials were used, and expected steps and methods were performed (documentation that procedure was followed as intended)
  - Assign tolerance limits to all of above

- Microbiological Surveillance Program
  - Environmental Control and/or Monitoring
    - Designated intervals
      - Cleaning
      - Air quality: in-process and at-rest, counts for viable and particulate
  - Microbiological culturing
    - Validate culture techniques for each cell/tissue type
    - Ongoing bioburden monitoring program
    - Assurance of adequacy: sensitivity and specificity
    - Recovery studies: Bacteriostasis and fungistasis testing
  - Validate processing methods to determine SAL (Sterility Assurance Level)
  - Assign tolerance limits to above

- Records
  - Requirements:
    - Documentation made concurrent with steps performed
    - Complete, accurate, legible, indelible, secure/protected when stored, available/ retrievable
    - Retention time (10 years, 30 years)
    - Ensure traceability of tissue at all steps (recovery to final disposition)
    - Confidentiality of donor and recipient information
  - Types, related to:
    - Donor: referral, consent, screening, "testing," family aftercare
    - Tissue: type, quantity, recovered, processed, preserved, packaged, labeled, culture results, "testing," release, storage/monitoring, distribution, recalls, final disposition
    - Personnel: qualifications, staff training, competency evaluations
    - Facility: cleaning and upkeep
    - Equipment cleaning, maintenance (calibration)
    - Complaints, SAEs, SARs
    - Policies, procedures, work instructions
    - Activity logs/databases, tracking trending/performance indicators
    - Electronic records!
      - Security administration, user responsibility and accountability, change control, validation vs verification

- Document Control
  - Require a uniform document format and identification system
    - Ensures consistency and avoids confusion
    - Require an approval system for documents prior to initial use
  - Periodic, planned, review system in place
    - Revisions must be easily identified (change/history file)
  - Maintenance of master files
    - Procedures, Forms
      - Labels
        - Type selected – appropriate adhesiveness/attachment to allograft during storage, distribution and use
        - Anonymous
      - Grafts made available (aka design control file)
  - Control system that ensures current form version is being used
    - Implementation often based on completion of staff training
  - Documentation "rules" (e.g., corrections, legible, indelible)
  - Signature log (staff initials, signatures used for signing documents)

- Quality Audits
  - Audit schedule/plan (frequency: periodic, annual, biennial, after change to a process or of personnel)
    - Internal
      - All operations; performed by qualified person(s) not performing or responsible for the work
    - External
      - Vendors/suppliers, contractors (e.g., entities that perform functions on your behalf)
    - Third party: required vs voluntary and "public trust"

- Investigations of Complaints, Departures, Deviations, Recalls
  - SAEs (Serious Adverse Events; aka Errors, Accidents) – Controllable, uncontrollable
    - Investigation, root cause analysis, identify solution(s), perform impact assessment, implement corrective and preventive action (CA/PA); monitor/verify effectiveness; management review
  - SARs (Serious Adverse Reactions) – Controllable, uncontrollable
    - Recognition (vigilance), reporting (surveillance), investigation, resolution, corrective action, report to competent authority
  - CAPA (Corrective and Preventive Action)

Note: Quality Program personnel are a demanding lot and they should be. Their systems checks are critical parts of operations as a whole. Verification that all critical steps are completed as expected and backed by proper documentation is a large task and is arguably required for consistent, successful outcomes. Their questioning during an audit or otherwise, when undertaken in an inquisitive but professional manner, should be portrayed and viewed as a point of learning for both parties.

Measuring Assurance and Expectations

Apply as necessary:

- *Qualification* = evaluation of equipment, reagents, materials, facility or personnel (specifications and checks)
- *Validation* = evaluation of a process such as treatments, testing, or claims (must show data)
- *Verification* = evidence the process was followed; evidence of process effectiveness (documentation and QC)

Determining Critical Points of Tissue or Cell Processing

Evaluate your process. What requires controls (qualification, validation, or verification)? This is used for your "master plan" that is described in your Quality Manual. For processing, this can involve these steps:

- Receipt of cells/tissues
  - o Specifications (use a checklist)
    - ▪ Package integrity (shipping and immediate container)
    - ▪ Transport conditions (temperature)
    - ▪ Time limitations
    - ▪ Labeling (identification code)
    - ▪ Other specifications (documents, blood samples, transport media, etc.)
  - o Documentation
    - ▪ Are specifications met? (accept or reject)
    - ▪ Date, time, and staff making determination (based on policies and procedures)
- Storage
  - o Specifications (temperature controls, as necessary)
- Facilities
  - o Environmental controls
    - ▪ Establish specifications/requirements, alert levels, limits
    - ▪ Identify critical areas (storage, processing/packaging, further and different storage) that require controls
    - ▪ Establish environment classification/grade requirements (e.g., A,B,C,D)
    - ▪ Develop maintenance and monitoring requirements (cleaning; temperature, humidity, air quality (viable and particulate)
- Supplies/Equipment/Instruments
  - o Material/reagents (establish specifications)
    - ▪ Determine adequacy (e.g., purity, grade, concentration) for intended use (e.g., packaging, labels, water, media, antibiotics, "consumables," etc.)
  - o Equipment
    - ▪ Storage devices, bandsaws, sterilization units, analyzers, "water systems," etc.
      - Perform installation qualification (IQ), operational qualification (OQ), performance qualification (PQ)
  - o Instruments/Tools
    - ▪ Drills, drill bits, brushes, scissors, scalpels, blades, forceps, calipers, etc.
      - Quality (grade) and other considerations (e.g., sterilized if multi-use)
- Treatments (in-process controls)
  - o Chemical, physical methods
    - ▪ Steps to inactivate or remove adventitious agents or "components"
      - Irradiation, disinfectants, antibiotics, surfactants, etc.
      - Decellularization, sonication, centrifugation, soaking, washing, cell separation, filtering, etc.
  - o Preservation methods
    - ▪ Freezing, lyophilization/dehydration, cryopreservation, vitrification, etc.
  - o Packaging method
    - ▪ Sealing process
- Testing (in-process controls to evaluate treatments and facility controls)
  - o Microbiological surveillance
    - ▪ Culturing methodologies (B&F testing, etc.); environmental monitoring; cleaning/sanitization of equipment, tools/instruments, and working surfaces
  - o Quality (physical or physiological specifications)
    - ▪ Residual moisture level, residual calcium level, BMP activity, tensile strength, cell viability, etc.
    - ▪ Claims such as: expiration date, sterility assurance level, antibiotic treatment, cytotoxicity, residuals, potency
- Staff
  - o Qualifications, Training, Competency

Future Considerations

Harmonization of tissue and cell banking regulations should be pursued by all relevant authorities since global distribution of these tissue and cell allografts is a reality and occurs regularly. This activity will only increase and expand as more countries successfully develop cell and tissue donation and banking programs and discover that surpluses can exist and more allografts can be shared. The health of patients in need of allografts should not be adversely affected by inadequate supply due to irregularities in regulations that do not, factually, protect them "more" and in reality, may cause an unintended consequence. This is a form of the "precautionary principle" gone awry. The intent of altruistic donation doesn't carry the burden of for whom the donation is used, but there is hope that the donation results in use. Expiring safe allografts that could have been used is not anyone's desire.

We know that successful quality programs can ensure a safe supply of tissue but this is not only applicable to tissue establishments. Our mission to provide patients with safe and effective allografts relies, in part, upon those who receive and use them because they play a critical role with an expectation to properly store, track, and prepare the cells and tissues we distribute to them. Quality concepts should be applied by the professionals who offer or manage tissue services [41]. Our service communities must work together to facilitate safe tissue handling.

Lessons we learn can be used to improve operations but they must also be shared with our professional colleagues via discussions, publications, presentations, and educational programs. You are encouraged to do just that.

Conclusion

An applicable, ending quote is one by Aldous Huxley (1894–1963), a prominent humanist who described society's misgivings:

> That men do not learn very much from the lessons of history is the most important of all the lessons of history.

In some ways, this, unfortunately, has proved to continue to be a true statement, however, public expectation of the safety of tissues and cells is high and we must not undermine this hope. In our profession, people can be intimately and profoundly affected by our services. This demands that sufficient quality measures be developed and implemented and operations monitored so we ensure stakeholder satisfaction, which includes our own. In general, the donation opportunity must be offered (a public right), it should successfully be carried out (correct moral conduct), and tissue provided for clinical use is expected to be safe (do no harm). By successfully applying this simple overview to tissue donation for transplantation, we can maintain the public's trust. Establishing and following well-designed quality systems and using the lessons we learn from applying this type of structure to tissue banking will ultimately meet expectations of all internal and external participants. Tissue banking has evolved tremendously in the past half century and has amassed a proud and prestigious safety record, but we must strive to always seek improvements and this can be reached best by cultivating an attitude of "quality" in everyday work and life. Quality must be habit.

A Note from the Author

The author wishes to pay tribute and extend heart-felt thanks to Ms. Deirdre Fehily. Tissue banking professionals are indebted to Deirdre for her exhaustive work in the area of developing and promoting quality concepts throughout cell and tissue banking operations in Europe. We should all be grateful for her numerous contributions that have positively impacted scores of unknown recipients and helped to make tissue banking and transplantation safe. A chapter about quality in tissue banking invariably includes an attribution to work in which she was involved.

References

1. Juran JM (1986) The quality trilogy: a universal approach to managing for quality. Qual Prog 19(8):19–24, Aug
2. Deming WE (2000) The new economics for industry, government, education, 2nd edn. Center for Advanced Engineering Study, Massachusetts Institute of Technology Press, Massachusetts, Aug 2000
3. Deming WE (1986) Out of the crisis. Massachusetts Institute of Technology Press, Massachusetts
4. Cortada JW, Woods JA (eds) (1998) The six values of a quality culture. The Quality Yearbook, McGraw-Hill
5. Mowe J (ed) (1984) Standards for tissue banking. American Association of Tissue Banks, Arlington, VA
6. Blood Action Plan (2001) Food and Drug Administration, Rockville, MD. http://www.hhs.gov/bloodsafety/BloodActionPlan.pdf. Accessed 30 May 2009
7. Tissue Action Plan (2005) Food and Drug Administration, Rockville, MD. http://www.fda.gov/BiologicsBloodVaccines/TissueTissueProducts/RegulationofTissues/ucm136967.htm. Accessed 30 May 2009
8. The ISO 9000 family, global management standards. http://www.iso.org/ Accessed 30 May 2009
9. Directive 2004/23/EC of the European Parliament and of the Council of 31 March 2004 on setting standards of quality and safety for the donation, procurement, testing, processing, preservation, storage and distribution of human tissues and cells. http://eurlex.europa.eu/LexUriServ/LexUriServ.do?uri=OJ:L:2004:102:0048:0058:EN:PDF. Accessed 30 May 2009
10. U.S. Department of Health and Human Services, Food and Drug Administration, 21 CFR Parts 207, 807, and 1271; Human Cells, Tissues, and Cellular and Tissue-Based Products; Establishment Registration and Listing; Final rule. Federal Register: Jan 19 2001 (Volume 66, Number 13); http://frwebgate.access.gpo.gov/cgibin/getdoc.cgi?dbname=2001_register&docid=fr19ja01-4.pdf. Accessed 30 May 2009
11. Therapeutic Goods Administration, Australian Code of Good Manufacturing Practice – Human Blood and Tissues, 24 Aug 2000; http://www.tga.gov.au/manuf/gmpbltic.pdf. Accessed 30 May 2009
12. http://www.tga.gov.au/regreform/index.htm. Accessed 30 May 2009

13. UK Dept of Health, A Code of Practice for Tissue Banks (2001) http://www.dh.gov.uk/en/Publicationsandstatistics/Publications/PublicationsPolicyAndGuidance/DH_4006116. Accessed 30 May 2009
14. Commission Directive 2003/94/EC of 8 October 2003 laying down the principles and guidelines of good manufacturing practice in respect of medicinal products for human use and investigational medicinal products for human use. http://ec.europa.eu/enterprise/pharmaceuticals/eudralex/vol1_en.htm. Accessed 31 May 2009
15. Directive 2001/83/EC of the European Parliament and of the Council of 6 November 2001 on the Community code relating to medicinal products for human use (Consolidated version: 30/12/2008), http://ec.europa.eu/enterprise/pharmaceuticals/eudralex/vol1_en.htm. Accessed 31 May 2009
16. http://ec.europa.eu/enterprise/pharmaceuticals/eudralex/vol4_en.htm. Accessed 30 May 2009
17. http://www.coe.int/t/dg3/health/Source/GuideSecurity2_en.pdf. Accessed 30 May, 2009
18. http://www.who.int/transplantation/AM-SafetyEssential%20HCTT.pdf. Accessed 30 May 2009
19. http://www.who.int/transplantation/AM-HCTTServices.pdf. Accessed 30 May 2009
20. U.S. Department of Health and Human Services, Food and Drug Administration, 21 CFR Parts 16, 1270, and 1271; Current Good Tissue Practice for Human Cell, Tissue, and Cellular and Tissue-Based Product Establishments; Inspection and Enforcement; Final Rule – 11/24/2004 (CFR Volume 69, Number 226); http://frwebgate.access.gpo.gov/cgibin/getdoc.cgi?dbname=2004_register&docid=fr24no04-9.pdf. Accessed 30 May 2009
21. U.S. Department of Health and Human Services, Food and Drug Administration, Draft Guidance for Industry: Current Good Tissue Practice (CGTP) and Additional Requirements for Manufacturers of Human Cells, Tissues, and Cellular and Tissue-Based Products (HCT/Ps) – January 2009 http://www.fda.gov/BiologicsBloodVaccines/GuidanceComplianceRegulatoryInformation/Guidances/Tissue/ucm062693.htm. Accessed 30 May 2009
22. U.S. Department of Health and Human Services, Food and Drug Administration, Guidance for Industry: Validation of Procedures for Processing of Human Tissues Intended for Transplantation, Final Guidance, March 2002, http://www.fda.gov/BiologicsBloodVaccines/GuidanceComplianceRegulatoryInformation/Guidances/Tissue/ucm073429.htm. Accessed 30 May 2009
23. Update: allograft-associated bacterial infections – United States (2002) MMWR 51(10). http://www.cdc.gov/mmwr/preview/mmwrhtml/mm5110a2.htm. Accessed 30 May 2009
24. Invasive *Streptococcus pyogenes* After Allograft Implantation – Colorado (2003), MMWR 52(48), Centers for Disease Control and Prevention (CDC), 12/12/2003
25. Kainer MA et al (2004) Clostridium infections associated with musculoskeletal-tissue allografts. N Engl J Med 350:2564–2571 http://content.nejm.org/cgi/content/abstract/350/25/2564. Accessed 30 May 2009
26. Commission Directive 2006/86/EC of 24 October 2006 implementing Directive 2004/23/EC of the European Parliament and of the Council as regards traceability requirements, notification of serious adverse reactions and events and certain technical requirements for the coding, processing, preservation, storage and distribution of human tissues and cells. http://eurlex.europa.eu/LexUriServ/LexUriServ.do?uri=OJ:L:2006:294:0032:0050:EN:PDF Accessed 30 May 2009
27. Commission Directive 2006/17/EC of 8 February 2006 implementing Directive 2004/23/EC of the European Parliament and of the Council as regards certain technical requirements for the donation, procurement and testing of human tissues and cells. http://eurlex.europa.eu/LexUriServ/site/en/oj/2006/l_038/l_03820060209en00400052.pdf. Accessed 30 May 2009
28. Regulation (EC) No 1394/2007 Of The European Parliament And Of The Council of 13 November 2007 on advanced therapy medicinal products and amending Directive 2001/83/EC and Regulation (EC) No 726/2004. http://ec.europa.eu/enterprise/pharmaceuticals/eudralex/vol1_en.htm Accessed 31 May 2009

29. IAEA *International Standards for Tissue Banks* (2005) http://www.int-tissuebank.com/bulletin/list.asp?bid=Documents&code=doc. Accessed 1 Jun 2009
30. IAEA *Radiation Sterilization of Tissue Allografts: Requirements for Validation and Routine Control – A Code of Practice* (2007) http://www.int-tissuebank.com/bulletin/list.asp?bid=Documents&code=doc. Accessed 1 Jun 2009
31. Safety of Human Cells, Tissues and Organs for Transplantation Regulations, *Canada Gazette*, Vol. 141, No. 13, June 27, 2007, Registration SOR/2007-118, June 7, 2007, Food and Drugs Act; http://www.gazette.gc.ca/archives/p2/2007/2007-06-27/html/sor-dors118-eng.html. Accessed 30 May 2009
32. Guidance Document for Cell, Tissue and Organ Establishments, Safety of Human Cells, Tissues and Organs for Transplantation, April 6th, 2009, Health Products and Food Branch, Health Canada (not yet available online as of May 30, 2009)
33. CAN/CSA *Z900.1-03* (2003) Cells, Tissues, and Organs for Transplantation and Assisted Reproduction: General Requirements, Jan 2003
34. CAN/CSA Z900.2.2-03 tissues for transplantation, Update No. 3, Mar 2007
35. European Quality System for Tissue Banking (EQSTB), http://ec.europa.eu/health/ph_projects/2003/action2/action2_2003_06_en.htm and http://ec.europa.eu/health/ph_projects/2003/action2/docs/2003_2_06_frep.pdf. Accessed 30 May 2009
36. Guide of Recommendations for Tissue Banking (2007) European Quality System for Tissue Banking (EQSTB)
37. Guide for Auditing Tissue Establishments (2007) European Quality System for Tissue Banking (EQSTB)
38. http://www.eustite.org/
39. http://www.eustite.org/files/InspectionGuidelines_Final%20Edit_July08.pdf. Accessed 30 May 2009
40. http://www.eustite.org/files/ToolsDeliverable10210508.pdf. Accessed 30 May 2009
41. Eisenbrey B, Eastlund T, Gottschall J (eds) (2008) Hospital tissue management – a practitioner's handbook, jointly published by AABB, EBAA, and AATB; AABB Press, Bethesda, Maryland

Chapter 13
IT System

Linda Lodge

An IT system can be defined as a technological solution comprising of the hardware, software and documentation components necessary to deliver an IT service. An IT system however is only part of the overall system. Operational processes and procedures need to interface with the IT system to allow the IT system to work as required and deliver the benefits.

In order to ensure an appropriate technology solution, operational requirements for all elements of the system must be clearly defined.

Failure to consider all IT system elements may lead to significant issues with usability, safety, quality, maintainability and cost. A lack of attention to detail at this stage may result in a technological problem rather than a solution. In other words rather than relieving an operational problem the IT system may simply change the problem or indeed make it worse.

What Are the Operational Requirements?

Consider the operational situation. What problem(s) require a solution? What benefits are sought? What processes is the IT system intended to provide or support e.g. donor management, process control, stock control, patient management, data analysis?

Ensuring that the operational requirements are fully understood and realistic, it is essential that those preparing the requirements have a good working knowledge of the business processes. It is important to get people from the right level, as those from the wrong level will bring assumptions rather than facts.

L. Lodge (✉)
Scottish National Blood Transfusion Service, Edinburgh EH17 7QT, Scotland
e-mail: Linda.Lodge@nhs.net

G. Galea (ed.), *Essentials of Tissue Banking*, DOI 10.1007/978-90-481-9142-0_13,

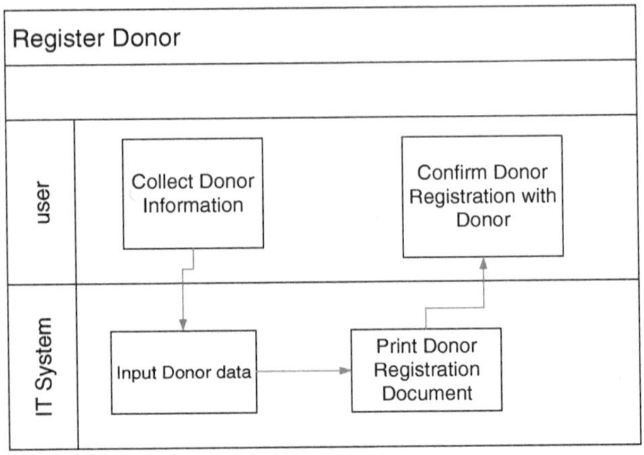

Fig. 13.1 High level process diagram

There are a number of sophisticated tools and techniques for determining, mapping and analysing operational business process and countless books available on the subject however in it's simplest form, a step through of every task and action performed, noting who does what, with what, when, where, why and how can prove to be an enlightening experience. Figure 13.1 shows a simple high level example of the Donor Registration Process. Detail can be added to achieve the required level of understanding for each process.

This exercise can be very useful and often highlights flaws in the existing process. It is possible that the solution for some problems may lie in a relatively simple change to operational process. Remember if an operational process does not work well on paper it will not work electronically. An IT system will not correct a flawed operational process.

Another potential benefit that can be gained from the operational business process mapping is the basic creation of Standard Operating Procedures (SOPs). As the processes are mapped it is possible to highlight the IT system user interfaces and develop the procedure that should be used by the user when operating the system.

Why Use an IT System?

Before deciding that an IT system is necessary consideration should be given to the following questions:

- Are actions continually repeated e.g. test result input?
- Does the process require complex algorithms e.g. virology repeat test management?

- Does a significant amount of data have to be kept and retrieved e.g. donor and patient details?
- Is all or part of the process prone to human error e.g. product code input?
- Are processes spread between departments or sites e.g. collection, testing, storage?

If the answer to one or more of these questions is "yes" then a properly defined and developed IT system could be part of an operational solution.

The Benefits of an IT System

An IT system has the ability to release significant operational business benefit. Consideration should be given to the operational situation and what benefits are desired. What are the priorities? Understanding the priorities will give a good indication of what is important in a suitable IT system. Potential benefits to a Tissue Bank operation may include some or all of the following:

- End to mountains of paper records, susceptible to loss or damage e.g. donor and patient records
- Rapid and easy access to structured donor and patient information at point of need
- Standardisation and efficiency of process e.g. multiple sites same process
- Reduction in transcription or translation errors e.g. use of barcode systems and electronic transfer of test results
- Security, access control to sensitive data e.g. access levels associated with system login
- Traceability e.g. donor to tissue to patient
- Ability to collect, index, search and analyse significant amounts of data e.g. operational and management reports
- Execution of complex protocols and algorithms e.g. virology repeat test management

In fact it is considered that the operations of a modern tissue establishment require most, if not all, of the above factors. They are essential for maintaining the safety of the grafts and to comply with modern legislation. For example it is a legal requirement (EU Directive 2004/23/EC) that traceability between donor and recipient is maintained. Handling of laboratory results using manual methods are significantly prone to errors, particularly when handling large numbers and the data is complex. Nowadays complex combined antigen/antibody tests are performed along with nucleic acid methodologies. Complex algorithms are needed to handle tissues and donors with repeat reactive results. The scope for error in such situations is high and some form of automated data transfer and handling is essential in my view.

Developing the Operational Requirements Document (ORD)

Once the reasons for considering an IT system solution are established and understood, the task of defining in sufficient detail the user operational requirements document/specification to be issued to potential suppliers can begin.

Requirements should be developed in terms of functional and non-functional requirements. Each requirement should be considered mandatory or non-mandatory. Careful consideration should be given to mandatory requirements as over specification in this area may exclude a supplier with a potentially excellent solution from making a proposal. It is often better to describe the "issues" and invite suggested solutions, the downside of this approach is that selection of the final solution can be more difficult to finalise as the proposed solutions may be very different.

- **Functional Requirements**
  Functional requirements are statements describing "what" the system must do. A typical example in Tissue Banking is "...must be able to register donor details". This requirement may be further expanded to detail what donor details are required e.g. Name, DOB, Sex etc. and may even state a specific data format e.g. Sex valid values = M or F. Consideration should be given to other system data dependencies and where possible data definitions and formats should be kept consistent e.g. date format.
- **Non-Functional Requirements**
  Non-functional requirements define how the system should operate. These are often forgotten as people get carried away with the possibility of new functionality. Non-functional requirements are however critical to a requirements specification if the system is to be usable within the Tissue Bank environment. An example of a non-functional requirement is that "...a label must be produced within 2 s of making the request". A usable label may become unusable if it takes 5 min to be produced from the printer. Typical non-functional requirements are concerned with areas such as system accessibility, scalability, availability and performance. It is also essential to consider operational growth and if there is no other substantiating data in my experience a reasonable assumption is between 4 and 8% per year over a 7 year period.

Technology Solutions

In response to the ORD a proposed IT system will consist of a collection of different technologies e.g. database management systems, data entry methods, data transfer protocols. Within these technology groups a number of options may exist. All or some may be advantageous in the Tissue Bank environment. Each option should be considered in terms of meeting quality and safety requirements, system and data integrity and interoperability with other systems and establishments.

Barcodes

Bar codes represent data in a machine readable format which can be read and decoded by a bar code scanner. The information is sent to the IT system where it is processed and recorded. Barcodes provide a secure method of data transfer due to the ability to make use of checksums and hierarchical data structures. Bar codes reduce the risk of data transcription errors by reducing the amount of human data keying required to enter the same data. They also reduce the risk of invalid data being entered into the system.

There are two main types of barcodes available, linear and 2-Dimensional. Within each type a number of bar code symbologies are available, each with their own advantages and disadvantages. It should be noted that the bar code symbology is only a vehicle to deliver a data structure. It is the ability to structure the system data that is important.

Linear Bar Codes

Linear symbologies are read by laser scanners using a single sweep of a beam of light from side to side across the barcode. Examples of linear symbologies in use are American Blood Commission (ABC) Codabar [1] and International Society for Blood Transfusion ISBT128 [2] (ICCBBA licenced [3]). ABC Codabar has limited use and does not provide total uniqueness when used for identification. ISBT128 is a more favoured symbology as it provides uniqueness and structures to incorporate varying types of data.

2-Dimensional Bar Codes

2-Dimensional codes e.g. DataMatrix2D [4] are seen as an efficient method of delivering large amounts of structured data securely. The most common symbologies use matrix codes featuring squares or dots on a grid and are smaller in size than conventional linear codes.

Radio Frequency Identification (RFID)

RFID is a technology used extensively for tracking the movement of packages with in the courier and retail industries. In healthcare this technology is used in some areas however care has to be taken in the selection and use due to the potential interference with other medical devices e.g. pacemakers [5]. While the potential advantages of RFID in Tissue Banking for the tracking of products is attractive the use of this technology in this environment is in its' infancy. More testing is required to understand the issues presented by storage at low temperature or in liquid nitrogen.

Electronic Data Interchange (EDI)

EDI is a set of standards for structuring information that is to be electronically exchanged between systems and businesses without human intervention. The EDI standards have been designed to be independent of software or communication technologies and messages can be transmitted using any sender/recipient agreed methodology e.g. modem, email. For Tissue Banking, EDI is commonly used for transmitting data between laboratory test equipment and the tissue management system.

Database Management Technology

A wide variety of relational database management systems are available e.g. Oracle [6], Microsoft SQL [7]. Database management systems are designed to internally manage the receipt, storage and retrieval of vast amounts of related data. Properly used this technology has the ability to ensure data integrity is maintained during operational use. It also has the ability to efficiently retrieve data ensuring operationally acceptable system performance. Some IT applications will only run on specific database technology while others can be run on a variety. It is also fair to say that the costs involved vary significantly between technologies, their implementation and support.

Selecting an IT System

In an ideal world there would be no constraints and selection of a system could be justified in terms of the "best fit" for the requirements. Unfortunately very few establishments are afforded that luxury. Considerations of time, effort and cost must also be justified. Therefore a well understood method of selection should be employed and agreed prior to the ORD being issued to potential suppliers. There is no absolute method for doing this however organisations usually employ a method of setting selection criteria and applying a weighted score to each, based on the relative importance to the business. The proposal with the highest score is deemed most suited to the needs of the business and therefore considered the preferred solution at this stage. This system is not infallible, how well have suppliers understood the requirements, how well have you understood the proposed solution and implementation, what other relevant information has it not been possible to factor into the selection criteria? I would suggest that this process is used to inform the decision but should not prevent other relevant factors being taken into consideration. For a complex set of requirements it is not unusual for the scoring process to be repeated once suppliers have had the opportunity to clarify some of the requirements and adjust their initial proposals accordingly.

During the assessment of a proposed system continually refer back to mandatory requirements. It is very easy to get carried away by slick demonstrations of nice to have functionality, especially where colourful graphics have been employed to show differing views of mundane data. The assumption is that "you too could have this with a simple press of a button", I have yet to see anything that simple with the ability to produce usable, meaningful results unless a lot of thought and effort has gone into defining the requirement, developing the functionality and collecting the data. While potential future requirements must also be considered beware of additional functionality, you don't want to be paying now for something that won't be used for 5 years.

Cost of an IT System

When considering and calculating the "true" cost of an IT system a number of elements need to be factored in. It is easy to miss (or ignore) significant costs that may put an unsustainable strain on a tight budget. Longer term it may reduce or prevent the expected return on investment being realised.

Costs should include all elements of system implementation. This should include all on-going support and maintenance costs estimated over a 5–7 year period. It is easier to think in terms of required functionality e.g. the ability to register a donor, and therefore define the IT application part of the system. Defining the hardware and operating platform requirements is a different matter. Some IT applications will only operate on specific platforms and this immediately ties the purchaser to the platform which may differ considerable from existing IT provision. Professional IT help, understanding the non-functional requirements and any existing IT infrastructure are essential.

If the IT application requires dedicated new hardware with a specialist specification the initial cost of implementation may be considerable. Add this to potentially specialist support and the on-going maintenance and very quickly the numbers begin to add up and make the IT application less attractive.

On-going support and maintenance costs must be fully explored. Operational requirements will change e.g. new tissues to be supplied, a new site opened, legislation changes, all of which have the potential to require changes to the IT system for both software and hardware elements. Understanding how and where costs may be incurred is vital when determining the "true" cost of an IT system. Some vendors will compare very favourably on price for the initial supply but will tie the IT application down so tightly that even relatively minor changes will have to involve the vendor and carry a large price tag once the IT application has been implemented.

It is worth considering when selecting a system how much change can be user controlled e.g. parameter driven test definition, versus what would have to be vendor dependent. Bare in mind however the more user control the greater the dependency on having sufficiently skilled staff to manage the system effectively.

Operational growth is another area often forgotten. Ideally the IT system should be sized from implementation to accommodate expected growth with little or no additional cost implications. An understanding of the point at which operational growth will incur substantial cost is an advantage as this can greatly influence the cost of an IT system over its' lifetime.

Failure to fully understand the implications in this area could result in a costly system implementation that fails to meet operational requirements in a very short period of time. Instead of supporting and benefiting operations the IT system could be holding it back unless money can be found to pay the high cost of change.

There is no right or wrong in this area and individual situations will ultimately determine what is done. The trick is to make sure that all elements have been factored in during the decision making process and the implication understood by all.

Implementing an IT System

The effort and time to implement a new IT system should not be underestimated but sadly all too often it is. This is an area that vendors and users alike tend towards, at best, optimistic, at worst impossible views and expectations. One of the reasons that so many IT implementations are reported to overrun is that initial expectations were unachievable from the outset. Even in a small Tissue Bank the effort can be considerable and effort is not necessarily directly (or solely) related to system size or complexity.

There is endless advice available on how to run a successful implementation in a variety of settings. Michael Hawksworth [8] however describes 6 fundamental steps that I think apply to any environment and fits nicely in the Tissue Bank environment.

- **Define success**
  What does success look and feel like? What are the deliverables from the implementation? What other dependencies exist? Remember the operational requirement statements. If 80% of tissues can be labelled via the system is that success? If a tissue from another establishment has to be relabelled to be accepted into your inventory is that success? It is important to establish IT and other non IT factors that may contribute to success and steps that may have to be taken to ensure success e.g. adoption of an international coding method e.g. ISBT 128 already proven within blood and tissue establishments, prior or in conjunction with the IT system implementation. Establishing these criteria at the start will determine the scope of the implementation and the dependencies that mean success.
- **Set priorities**
  There is rarely one goal or a simple problem to be solved. The solution will be required to solve a number of problems e.g. print labels, register donors, electronically accept test results. While different individuals may see different priorities it is wise to have overall agreement with regard to priority. Is the ability to print labels with in 2 s more important than the ability to correctly link donor to tissue?

By setting priorities implementation can stay focused on what is important and not waste time and money gaining an odd second in performance while forgetting to ensure all patient identifiers are included.

This approach may also show opportunities for phased implementation. If the most significant functionality can be implemented relatively quickly there is an opportunity to release system benefits sooner. Less important functionality can be added in future phases. This approach can help with the cost benefit analysis and reduces the pressure on operational staff trying to cope with a new system and maintain operational processing.

- **Avoid modifications**
  There are different levels of modification that may be requested during an implementation. While some low level modification e.g. parameter configuration is inevitable, fundamental modification to systems that require significant system redesign or source code changes should be avoided. If the right application has been selected redesign work should not be necessary. If a system can't link a tissue to a donor in a few steps, is it the right system for you? If the screen background is grey and you wanted blue how important is it?

  Changes at a fundamental design level are time consuming and costly and may affect the overall integrity of the system.

  Changes to system configuration are highly likely e.g. defining test codes however once the system has entered formal validation even these changes can have significant impact on time, effort and cost.
- **Prepare for change**
  While some people embrace change there are a large number who do not. It is essential that end users and managers who have to make the system work are involved from an early stage. If business process has to change to accommodate a new system then the earlier this is identified and communicated the better.

  Involve as many as possible from across the organisational hierarchy in defining requirements, validation and implementation planning therefore giving an opportunity to participate and take ownership of their processes. Let people see what is coming as soon as is practicable.
- **Gain executive support**
  If the people at the top are on board and support the implementation it will be easier to define success, get high level decisions, support during difficulties and access to the resource needed to successfully implement.

Money and People

Often neglected is the amount of time required from operational people e.g. departmental managers and users, during an implementation. Those who manage donor consent, process tissue, distribute tissue are best suited to define requirements in these areas and validate solutions and therefore should be consulted and involved throughout development and implementation. This can require significant effort, make sure enough is available otherwise time to implement will increase and the solution may not be the best it could be.

Who will manage the implementation? Vendors may offer the service but who will manage the operational side of things. If you do not have someone with the appropriate skill and experience then bringing someone in for the duration of the implementation could be money well spent.

Physical Implementation

It is good practise to employ a structured approach to project management and implementation. In Tissue Banking a risk based approach is necessary to provide a GxP delivery e.g. Projects In Controlled Environments II (PRINCE2 [9]) following Good Automated Manufacturing Practise (GAMP5 [10]) Guidelines. Relevant competent authorities will expect this level of control and structure to be demonstrated during any post implementation inspection.

System Validation

A major part of any implementation is system validation. When planning an implementation an estimated 40% of overall implementation effort should be allowed for validation.

Validation can be divided into different types of testing typically including functional testing, load and performance testing and user acceptance testing.

Functional Testing looks at the system requirements and tests the functionality to ensure that the ORD requirements have been satisfactorily met.

Load and Performance Testing applies operational load to the system in terms of number of users and volumes of transaction and measures performance to ensure that it is operationally fit for purpose.

Representatives from the business who will be using the system once implemented should conduct User Acceptance Testing. This testing tends to select potential "real life" scenarios which are run through the system and the results examined to ensure that they produce the expected outcomes.

Validation will continue post implementation as changes are sure to be required. It is essential from a GxP point of view that the validated state is maintained. If possible it is recommended that automated testing tools are used to create test plans and test scripts as these can then be easily maintained and reused throughout the system lifecycle.

User Training

While early exposure to a system is beneficial, formal training can only realistically start once the system is stable and representative of what the final product will look like.

It is essential that users are sufficiently trained and have confidence in the system and their ability to utilise the system in operation. If there are issues with user training and confidence early running of the system may be problematic with users making errors resulting in incorrect system usage. This can give an unrealistic view of the system.

Managing an IT System

How an IT system is managed will determine the impact on Tissue Bank operations and will in turn be determined by the expertise available to the Bank. If the Tissue Bank is large or part of a larger organisation it may have the ability to host the hardware and manage the system in-house. Consideration should be given to the effort and costs involved in terms of suitable accommodation for hardware and communication links, skilled staff and system availability e.g. if the system goes down for any reason over a weekend what would the impact be? In-house management needs to be able to cover all aspects of the operational need.

There are vendors who will offer a "hosted only" or a "fully managed service". The terms and support vary from vendor to vendor. Hosted only typically means that the vendor will "host" your application on their hardware and ensure that the hardware supports the requirement. Fully managed service means that the vendor will both host the hardware and "look after" the application with some kind of guarantee in terms of quality of service.

The costs of each option need to be compared against the resources and ability of the Tissue Bank.

It should be remembered however that even a fully managed service will require effort for Tissue Bank staff. Someone must manage the service contract and deal with any issues of service level performance. Also there are certain activities that may still be retained within the Tissue Bank or at least require attention from time to time e.g. new user account, password management, adding a new test code. It is important that a level of operational knowledge of the IT system is developed and retained because if the vendor or system changes for any reason the Tissue Bank may find that it no longer has access to significant business knowledge.

Conclusions

I hope to have shown the importance of planning and the operational requirements for establishing an IT system in a tissue bank. There is considerable thought that needs to go into assessing the requirements and how an IT system best supports them. Although there are financial implications and there has to be ongoing commitment to the IT system chosen to ensure that it remains functional and up to date, the benefits in terms of safety, GMP and traceability are very significant. A dedicated IT system should be on the requirement list of any modern tissue establishment.

References

1. American blood commission (CCBBA) now (ICCBBA, see Ref 3)
2. International Society of Blood Transfusion. www.isbt-web.org
3. International Council for Commonality in Blood Banking Automation. www.iccbba.org
4. Data Matrix Data Matrix was invented by RVSI/Acuity CiMatrix, who were acquired by Seimens AG in October, 2005 and Microscan in September 2008. www.microscan.com
5. van der Togt R et al Electromagnetic interference from Radio Frequency Identification inducing potentially hazardous Incidents in critical care medical equipment. www.jama.com
6. Oracle Corporation. www.oracle.com
7. Microsoft Corporation. www.microsoft.com
8. 6 Steps to ERP Implementation Success, Michael Hawksworth Jan 2007
9. PRINCE2 (PRojects IN Controlled Environments) is a process-based method for effective project management. Version 2. www.prince2.com
10. International Society for Pharmaceutical Engineering (ISPE) Good automated manufacturing practice version 5 (GAMP5). www.ispe.org

Part V
Legal and Ethical Environment

Chapter 14
Regulatory and Ethical Issues

David Pegg

Introduction

Almost all medical care involves interaction between patients and health care professionals: as a result, medical ethics are normally concerned with the rights of the patient and the consequent obligations placed on those providing medical care. However, the transfusion, implantation or grafting of human material involves another group – the donors of the cells, tissue or organ – whose role is humanitarian but non-professional. The discussion in this chapter will be very largely concerned with the rights of this third group and the way in which those rights create obligations for the rest of society, including health care professionals and even the patients who benefit from such treatment.

The nature and extent of human rights have evolved throughout history and are now embodied in The Universal Declaration of Human Rights (UDHR) that was issued by the United Nations on December 10th, 1948, just 60 years ago. In 1976 [1]. The International Covenant on Civil and Political Rights (which was based on the Universal Declaration of Human Rights) entered into force and all who have ratified this treaty are now bound by it [2]. Article 12 of the UDHR protects the privacy, autonomy and dignity of the individual. Article 25 stipulates access to medical care in order to ensure health and well-being. Treatments that involve the implantation, transfusion or grafting of cells, tissues and organs are now able to relieve many medical conditions and such treatment has become an intrinsic part of modern medical technology. While giving patients the right to expect this form of treatment in appropriate circumstances, the UDHR also insists that the participation of the donor must be freely given and not involve interference or pressure of any kind.

D. Pegg (✉)
Department of Biology (Area 14), University of York,
York YO10 5YW, UK
e-mail: dep1@york.ac.uk

G. Galea (ed.), *Essentials of Tissue Banking*, DOI 10.1007/978-90-481-9142-0_14,

In this chapter I shall argue that there are three fundamental ethical requirements that derive from the human rights of the donor and the recipient: the use itself must be ethical; there must be valid consent to donation, actual use and any consequences of donation; and there may be no harm or damage to, or degradation of the donor.

Ethical Uses of Human Tissues: Transplantation and Transfusion

In western society the therapeutic use of human tissue is, in itself, generally accepted and so is the use of tissues for research or teaching. However, there can be individual objections that usually derive from religious beliefs such as those held by Jehovah's Witnesses. There has been some debate within the Islamic community focusing on concern that the donation process necessarily involves disrespect to the donor's body. But the consensus among Islamic scholars is that organ donation is not only permissible but is in fact desirable. Others may object on a purely personal basis and although this is probably uncommon the right of individuals not to participate in tissue transfer must be, and is, protected. Other potential uses of human material, for food, clothing, ornament or entertainment, are not generally accepted in western society. The borderline between entertainment and education can be difficult to define precisely: the controversial German anatomist, Gunther von Hagens [3] has dissected human corpses for public display: his audience is the general public, not medical students and whether this transgresses the boundary between education and entertainment is debatable. Many will find the practice at least distasteful but it should be noted that the new legal framework does allow for the possibility of public display of human bodies or body parts [See Human Tissue Act 2004, Schedule 1, Part 1, Purpose 5 is Public Display].

The Crucial Nature of Consent

Consent is at the centre of ethical considerations of the transfusion or grafting of human material. This centrality is derived from the concept of "the inherent dignity and the equal and inalienable rights of all members of the human family. (UDHR)" Human beings are not to be subjected to procedures without having given their consent. According to the Shorter Oxford Dictionary, the word "consent" signifies a voluntary agreement to, or acquiescence in, what another proposes or desires. Thus, the giving of consent is a deliberate voluntary act based on an understanding the issues involved. The writers of the Nuffield Council on Bioethics report [4] argued that, because description can never be exhaustive and may be incompletely understood, the terms "informed consent" and "fully informed consent" should be avoided. To quote,

> The ethically significant requirement is not that consent be complete but that it be genuine. Ensuring that consent is genuine is mainly a matter of care in detecting and eliminating lack of consent. Both in law and in ethics, consent requirements are not met whenever anything rebuts or defeats the presumption of consent. The ascription of consent is defeasible: the presumption of consent can be defeated by any of numerous circumstances, including violence, coercion, deception, manipulation, tendentious misdescription of action, lack of disclosure of material facts or of conflicts of interest and the like.

The North American so-called "objective standard" of informed consent was rejected by the House of Lords [5]. But, granting that complete understanding is unattainable, it is nevertheless a requirement that the person giving consent has sufficient understanding for all those involved to be satisfied that the decision was soundly based. In this chapter the term "valid consent" will be used to describe such consent. The essence of valid consent is that there may be no coercion or manipulation of the person giving consent, and the person seeking consent must provide a description that honestly attempts to meet the needs of the person who is being asked to make the decision.

The Question of Ownership

Many situations that require consent do so because the person who is asked to give their consent is the owner of the material or the process involved. Is ownership the basis of consent in tissue grafting? There is a variety of attitudes to these matters in western society as illustrated by the several letters to the Editor of the Guardian newspaper that were printed on July 10th 1999 [6] under the heading "My body belongs to me." Some correspondents regarded their body as their personal property to be dealt with in the manner of their estate ... according to their will. These wishes may include specifying a particular use of their tissues, or some characteristics (religion; race) of the recipient of a graft. Others argue that when they die and their body is no longer of any use to them, that it should become the property of the community to be used for the greatest good of the greatest number – a sort of "utilitarian nationalization of corpses". However, the traditional legal position has been that there can be no property rights in bodies or body parts and that therefore there can be no requirements that are based on ownership. Confusingly, there have been some legal decisions in which it was held that when tissue had been processed in some significant way it then became the property of the person or organization that carried out the processing. It is difficult to define an ethical basis for that view and it would create all sorts of anomalies; an unprocessed femoral head would not be the property of the Tissue Bank but a cryopreserved heart valve would. So would tissue processed for histological examination, even though it has been conceded that such tissue is not the property of the Pathology Department that processed it. But in fact there is no need to invoke the concept of property; there is a distinction between ownership and legal possession. (I am in legal possession of my library book but it is the property of the library). Tissue Banks are in legal possession of donated tissue but are not, and do not need to be, its owner. The requirement for consent comes from Human Rights considerations not ownership.

Living donors

Donors may be alive at the time of donation or they may have died. Tissue from living donors may be incidental to a surgical procedure carried out for the benefit of the donor, such as the replacement of a hip joint by a mechanical prosthesis which incidentally makes the femoral head available for grafting. Alternatively the donation may have no therapeutic benefit to the donor but be a voluntary and deliberate gift, usually to a relative, for example of a kidney or part of a liver. Living donors of either type give consent for the donation and the specified use of their own tissues. As we have discussed, valid consent is required for three separate aspects of tissue donation; the act of donation itself, the use to which the tissue will be put and any consequences of donation. The Human Tissue Authority independently scrutinises all donations from living donors in order to ensure that donors are fully aware of any risks they may be exposing themselves to and to ensure that there is no coercion.

Consent to Donation

Valid consent to donation requires a complete disclosure of the donation process and of any risks or ill effects to the donor that could be associated with the act of donation. This is particularly important if the process of donation is of no medical benefit to the donor, as is the gift of a kidney.

Consent to Use

Valid consent to use requires a description of those features of possible uses that the subject needs to reach a decision. A simple description should always be provided of any of the following general uses that are envisaged:

- direct therapeutic use as a graft
- ethically-approved, academic medical research
- use by commercial medical entities

A leaflet giving more detailed descriptions may be available and should be offered to the donor. Different donors will need differing amounts and types of detail to reach a decision and the donors themselves should be helped to grade the amount of detail that they receive. Sometimes there is no way of knowing at the time of donation exactly what donated tissue may eventually be used for or when it will be used. Where this is the case, donors should be made aware. If consent is to be sought for research or commercial use particular care should be taken to provide sufficient information. In particular, blanket consent to unspecified research is not considered to constitute valid consent [4].

Consent to the Consequences of Donation

Valid consent also requires that the donor understands and accepts any possible consequences of donation. These include the discovery of infections that could have implications both for the donor and others, and an awareness of the inevitability of eventual disposal of all non-transplanted tissue. It is conceivable that a donor may ask not to be told the results of virological tests, and consequently may be allowed to make that stipulation at the time of donation. However, that would have implications for third parties outside the donor/transplanter relationship and, with the exception of the collection of anonymised tissue for epidemiological studies and the like, the duty of care to the family of the donor requires that agreeing to be informed should be a condition of donation. This may deny some potential donors the opportunity to donate.

The policy of the organization concerning the disclosure to graft recipients of the identity of the donor, and vice versa, should be explained. This assumes greater importance if the donor is deceased and is discussed in more detail below.

Situations can arise in which it is impossible to obtain valid consent to donation in sufficient time, which leads to the question, "Can retrospective consent ever be valid?" Where the technical processes involve no more than ensuring that tissue that had already been removed in the patient's interest would remain suitable for donation if the donor so wished when asked later, then it is difficult to see why that would not be acceptable. Examples include the collection of cord blood after delivery of a placenta or placing an excised femoral head in a sterile container. In such circumstances donation would not be considered to have taken place until valid consent was given; if consent was withheld, then the placenta and cord blood or the femoral head would be disposed of in the normal way. But prior consent, with adequate time for consideration should be the norm and there should always have been a good and sufficient reason for not obtaining consent in advance and such reasons should be recorded at the time.

Withdrawal of Consent

In many situations it is normal to allow a person an opportunity to withdraw their consent – to provide time for "second thoughts" or a so-called "cooling off period". The Convention on Human Rights and Biomedicine stipulates, at Article 5, that any intervention in the health field may only be carried out with the person's consent and *that person may freely withdraw consent at any time* [7]. This cannot be applied literally to tissue banking, particularly in view of the differing timescales of donation and use. It is suggested that it is both logical and ethical to allow consent to donation to be withdrawn up to the moment when the donation process commences, but not thereafter. A gift is a gift and the donor should accept that gifts cannot be reclaimed. The question then arises,"Can consent for use be withdrawn some time after the donation?" It is obvious that consent cannot be withdrawn after transplantation of a tissue graft or when a recipient has commenced treatment in preparation for that specific graft, because that would interfere with the autonomy of an innocent third party. Nor can there be any basis in property rights (*vide supra*) for the donor to withdraw permission for use of tissue that has been donated to, and is held by, a tissue bank. It seems inescapable that permission for use cannot be withdrawn once donation has taken place. However, the existence of accurate tissue tracking arrangements means that it would not be difficult to comply with a request to withdraw such stored tissue and it would be wise, without conceding any rights, for tissue banks to comply with any such request.

Deceased Donors

The three elements of valid consent that apply to living donors also apply to cadaver donation but with the fundamental difference that consent is being sought from a relative with only indirect access to the wishes of the deceased donor. The crucial question is whether the relatives are acting independently of the deceased or are acting as proxy for him. It was the clear intention of both the 1961 legislators and the 2004/2006 legislators of the current Human Tissue Acts, to give primacy to the donor's wishes and in practice it is rare for a family to object when there is clear evidence that the deceased had chosen to donate. But most people do have powerful feelings about the bodies of their relatives and the process of obtaining valid consent to cadaver donation is necessarily and properly influenced by the attitude of the acutely bereaved relatives. However, we must also be concerned that, when the donor was a legally competent adult, his own wishes should receive proper attention. It is true that a donor may not always fully understand what is involved in donation when he completes a donor card or puts his name on the donor register. He may not have discussed these matters within the family and nor may he have considered how his relatives will react. The question then arises, "Do these considerations affect the validity of consent?" In as much as the information was available to the donor if he chose to seek it the voluntary registration as a potential donor would be valid. Although only about 20% of potential donors carry donor cards, the card and the register have great value in making the wishes of the deceased clear to the relatives.

The structure of the interview in which consent is sought should draw the attention of the relatives to the wishes of the deceased if those are known or can be presumed. It is civilised and ethically sound to respect the last wishes of a fellow human being and it is also true that some people object strongly to the possibility of their relatives over-riding their wishes after death. Nevertheless, in situations where the wishes of the surviving relatives and the deceased are in tension and it is impossible to satisfy both, then the balance should be allowed to tip in favour of those who are still alive, always providing that the relatives concerned have addressed the issue squarely. It is also important to note that without the family's cooperation it is not possible to obtain a reliable medical and behavioural history.

The interview procedures in current use are very detailed and cannot reasonably be criticised for withholding or limiting the information that is available to relatives. On the other hand it may be asked whether so much detail is always necessary to establish valid consent. One wonders just how much information relatives actually absorb, particularly concerning possible research uses of donated tissue when the justification for the request may be difficult to explain in non-technical terms. The demand from potential research users of tissue is already considerable, is increasing and is likely to increase further, but at the present time, some up to 50% of relatives do refuse consent for tissue to be used for research. In some cases a Research Ethics Committee may have specified the extent of information to be given. It is crucial that information be put across in a way that is understood and that each interview is tailored to the needs of the individual relative. As with living donors, it seems good practice to provide some of the more detailed information in a leaflet.

Accurate recording of consent is particularly important for cadaver donation since the interview is not usually face-to-face but over the telephone. Voice recording does not affect the validity of consent, which depends solely on the content of the discussion between the coordinator and the person being interviewed. Recordings can be valuable to both parties if subsequently there were to be uncertainty as to what had been agreed. Training of the interviewer is very important and signed "consent forms" that have been obtained by untrained hospital staff do not constitute valid consent; they are dangerous because they may be taken as authorisation for tissue retrieval by pathologists and mortuary staff.

Disposal of Tissue Obtained from Deceased Donors

Relatives are often concerned about what will be returned to them for burial and this should be clearly explained. It must also be explained that all tissue that is not transplanted, whether surplus tissue from processing, outdated unused tissue or tissue that has been used for research, cannot be returned but will be disposed of. Where relevant, the relatives should also be aware that the processes used in tissue banking may necessitate the disposal of some tissue on several occasions over a prolonged time period. At the present time such tissue is disposed of by incineration as clinical waste and it may be that the practical realities of such disposal methods do not meet the concerns of some relatives. The term "respectful disposal" is sometimes used in this connection but the precise meaning is unclear. It is axiomatic that all phases of handling donated tissue should be respectful and that includes eventual disposal; the problem is that attaching the word "respectful" to the word "disposal" may, for some, imply solemn ceremony which, to others, seems both unnecessary and inappropriate. In view of its different implication, use of the word "respectful" is perhaps best avoided in this context.

The Passage of Information Between Recipient and Donor

Blood transfusion is anonymous and so is most tissue donation. If the donor is deceased it is common practice to send a standard letter to the relatives several days following donation, thanking them for the donation that was made altruistically for the benefit of others. However, some of those involved, in organ transplantation from deceased donors in particular, advocate giving limited information about the recipient to the donor family: they argue that this gives a measure of consolation to the donor's relatives, helps the grieving relatives of the deceased donor and publicises the benefits of transplantation to the general population. If information is to pass between recipient and donor, this will be an important consequence of donation that should be declared at the time of obtaining consent. But this writer would argue against such action. First, the donor family has no right to that information: the donor family were acting as proxys for the donor, presenting the donor's wishes and not making an independent decision. In fact, giving such information can create feelings of debt, obligation and guilt in the recipient and the donor family may want to make and maintain contact with the recipient. The fact that a part of the donor is still alive in the recipient's body may actually make it more difficult to come to terms with the death of the donor. There can also be problems in the flow of information in the reverse direction – recipient to donor family. Accurate information may not always be available, some transplants are unsuccessful and there is the practical problem that tissue from a single donor may be used for numerous recipients over a relatively long time. All these problems can be avoided by retaining the Blood Service philosophy of anonymous, altruistic donation for the general good of the community [8]. But if this policy should change, and recipient details be passed to the donor family or vice versa, then those giving consent and those receiving grafts should be made aware that they may receive or be asked to provide such information at a later date.

The Law

Tissue Banking in the UK is controlled by The Human Tissue Act 2004. and Human Tissue (Scotland) Act 2006 [9, 10]. These acts provide for the removal, storage or use of human tissues subject to two conditions: that the purpose is a scheduled purpose; and that appropriate consent (or authorization in Scotland) has been given. The scheduled purposes are: determining the cause of death; establishing after a person's death the efficacy of any drug or other treatment administered to him; obtaining scientific or medical information about a living or deceased person which may be relevant to any other person (including a future person); public display; anatomical examination; research in connection with disorders, or the functioning, of the human body; and transplantation.

The Westminster Act carefully defines the appropriate persons to give consent as follows:

Alive: His/her consent.

Deceased: If he/she had made a decision before death, his/her consent. Otherwise, the consent of a person nominated by the donor during life, before witnesses, to deal with this question, but only if it is reasonably practicable to communicate with this person in the available time. Otherwise, the consent of a person in a qualifying relationship.

Incompetent: If he/she had made a decision while competent, his/her consent. Otherwise, only if it is permitted under Regulations to be made by the Secretary of State

Presumed: If human tissue that came from a living person who cannot be traced or who has been consulted but will not make a decision, and if there is no reason to believe that the person had refused consent, and if the proposed use of the tissue is to provide medical information that is in the interests of another person or persons, then the Human Tissue Authority may stipulate that consent be presumed.

The concept of "presumed consent" is discussed below. Appropriate persons to give consent for child donors are also specified. Also, the act provides a ranked list of qualifying relationships, as follows:

(a) spouse or partner;
(b) parent or child;
(c) brother or sister;
(d) grandparent or grandchild;
(e) child of a person falling within paragraph (c);
(f) stepfather or stepmother;
(g) half-brother or half-sister;
(h) friend of longstanding.

Relationships in the same category have equal ranking. Consent is to be obtained from the person who has the highest ranking. If the relationship of each of two or more persons is accorded equal ranking then it is sufficient to obtain the consent of any of them. But any person's relationship shall be left out of account if he/she does not wish to deal with the issue of consent, or he/she is not able to deal with that issue, or it is not possible to communicate with him/her in the time available. The Scottish legislation is simpler, though probably no different in intent or practice.

The Westminster Act specifies offences and penalties. To proceed with a scheduled purpose in the absence of consent (unless the person concerned reasonably believed that consent had been given) or to proceed with any purpose that is not scheduled is an offence. Commercial dealings in human material for transplantation are prohibited. The penalty may be a fine or up to 3 years imprisonment or both. Existing holdings of human tissue are exempt from the need for consent. There are provisions for dealing with museum collections of anatomical, pathological or archaeological specimens and religious relics and other specimens that are more than 100 years old. Most importantly the Act set up the Human Tissue Authority to regulate the use of relevant human tissues for all the scheduled purposes and the HTA covers Scotland as well as the rest of the UK for licencing purposes. However, blood and anything derived from blood is excluded. The Authority is issuing Codes of Practice and is inspecting and licensing premises. One of the Codes of Practice deals with the central matter of consent in a practical way. But it does specifically state that "giving consent is a positive act. The absence of refusal is not evidence of consent" (para 17). This is interesting in the context of the presumed consent debate.

Presumed Consent

The Department of Health in the UK has recently conducted a survey of opinion concerning "presumed consent". The argument is that there is a gross shortage of organs for transplantation (true) whereas there probably are significant numbers of potential donors who never become actual donors because they do not, for one reason or another, sign the present opt-in donor register. One proposed action would be to replace the opt-in register by an opt-out register. According to this philosophy, consent is presumed if it has not been cancelled by an entry in the opt-out register. It will be immediately apparent that this would conflict with para 17 of the Code of Practice on Consent. In fact, the term "presumed consent" is an oxymoron since, according to the Shorter Oxford Dictionary, "consent" is a voluntary agreement to or acquiescence in what another proposes or desires. Since consent is a voluntary act it cannot be presumed: there must always be a deliberate affirmative action by the person giving consent. In fact, presumed consent is not consent at all! But a more important point is that the role of consent in medical procedures has received exhaustive ethical analysis, perhaps most precisely in relation to human tissues, by the Nuffield Council on Bioethics referred to above [4]. There, it is argued that the fundamental basis of the requirement for consent is the adage *volenti non fit iniuria* – no wrong is done to one who is willing. The authors argue that genuine consent is restricted to an act that is described in a specific way. Thus, there must be an exchange of information that meets the requirements of the potential donor or his relatives in order to establish that valid consent has been given. The absence of objection can never guarantee that a potential donor is indeed willing to be an actual donor. There may be many reasons for failure to register an opt-out, any of which would render the presumption of consent unsafe: ignorance of the issues; misconception regarding the issues; inability to reach a conclusion; reluctance to cooperate with Government schemes in general; laziness; forgetfulness; distraction. In fact, the current trend is to extend rather than limit the rights of the individual to make decisions that affect themselves; opting out would be a step in the reverse direction.

An important practical point is that there is no proof that adopting an opt-out system would increase the supply of donors; if the public reaction was negative it could even reduce the supply. Evidence from other countries is unconvincing because there are always other confounding factors such as the number and mode of operation of transplant coordinators. In practice an opt-out scheme would actually operate as an opt-in system because the relatives would always have to be consulted for safety reasons and if the relatives objected the donation would not take place. In fact the experience in other European countries suggests that the adoption of opting out in itself, has little effect on actual rates of organ donation. On the other hand is strong evidence that the provision of an extensive and efficient system of organ retrieval has a major effect.

It is concluded here that changing to an opt-out system would be ethically questionable and an unnecessary affront to the rights of the individual. Common sense suggests that an improved and extended transplant coordinator system and altered attitudes in hospitals, as proposed in Recommendations 4 and 9 of the first report of the Department of Health's Organ Donation Taskforce [11], would do much to increase the supply of tissues and organs for transplantation. To be effective this would, of course, require considerable additional expenditure. The working party has now produced its report which thankfully recommends the retention of the present opt-in system . The Government has accepted this albeit with the fall-back position hints from the Prime Minister that if the supply of organs for transplantation does not increase they may he will want to look again at an opt-out scheme.

The humane resolution of the problem of an inadequate supply of tissues and organs for transplantation may come from technical developments in tissue engineering and the development of artificial organs and tissues. Improvement in the supply of living, human material requires, above all, improved public education.

Acknowledgement The author is a consultant to the Blood Service and the current Chair of the Standing Advisory Committee on Tissues. He chaired the discussions that took place between members of SACTSC Sub-group on Consent to Tissue Donation in 2003. The participants included Prof Brenda Almond (Social Values Research Centre, University of Hull), Ms Jane Griffiths (Donor Transplant Co-ordinator, North Thames), Elizabeth Melville (Tissue Coordinator, SNBTS, Edinburgh), Ms Jan Purkis (Tissue Co-ordinator, NBS Brentwood), Dr Fiona Regan (Consultant Physician, NBS Colindale and BBMR), Dr Francis Rushambuza (Consultant Physician, NBS Liverpool & Wrexham Tissue Banks), Dr Chris Womack (Consultant Pathologist, Tissue Bank, Peterborough Hospital).

I am grateful to all members of this group. This chapter draws heavily on the report of that group to the Blood Service. Its content is, however, entirely my responsibility. I also wish to thank Dr Asad Khan for his advice concerning the Islamic position on organ donation and transplantation.

References

1. The Universal Declaration of Human Rights, United Nations, 10 Dec, 1948. Available at http://www.un.org/Overview/rights.html
2. The International Covenant on Civil and Political Rights, 23 Mar 1976. Available at http://www.unhchr.ch/html/menu3/b/a_ccpr.htm
3. Gunther von Hagen. See this link for a 'More4' documentary. http://www.channel4.com/more4/documentaries/doc-feature.jsp?id=57
4. 'Human Tissue. Ethical and Legal issues' ISBN 0 9522701 10. See particularly paragraphs 6.19 and 6.20
5. Jones MA (1999) Informed consent and other fairy stories. Med Law Rev 7:103–134
6. Letters to the Editor of the Guardian newspaper on July 10th 1999
7. Kennedy Institute of Ethics Journal 7.3:259–276, Available at http://muse.jhu.edu/journals/kennedy_institute_of_ethics_journal/v007/7.3dommel.html, 1997
8. Titmus RM (1970) The gift relationship. From human blood to social policy. Allen and Unwin, London
9. Human Tissue Act (2004) The Stationery Office
10. Human Tissue (Scotland Act 2006) The Stationery Office
11. Reports of the Organ Donation Taskforce, Available at this link http://www.dh.gov.uk/en/Healthcare/Secondarycare/Transplantation/Organdonation/index.htm

Index

G. Galea (ed.), *Essentials of Tissue Banking*, DOI 10.1007/978-90-481-9142-0,

FSC
www.fsc.org
MIX
Papier | Fördert gute Waldnutzung
FSC® C083411

Kolibri 360 GmbH
Ferdinand-Jühlke-Straße 7
99095 Erfurt, Deutschland
produktsicherheit@kolibri360.de